The Marsh Builders

The Marsh Builders

The Fight for Clean Water, Wetlands, and Wildlife

SHARON LEVY

OXFORD
UNIVERSITY PRESS

OXFORD
UNIVERSITY PRESS

Oxford University Press is a department of the University of Oxford. It furthers the University's objective of excellence in research, scholarship, and education by publishing worldwide. Oxford is a registered trade mark of Oxford University Press in the UK and certain other countries.

Published in the United States of America by Oxford University Press
198 Madison Avenue, New York, NY 10016, United States of America.

Library of Congress Cataloging-in-Publication Data
Names: Levy, Sharon, 1959– author.
Title: The marsh builders : the fight for clean water, wetlands, and wildlife / Sharon Levy.
Description: New York, NY : Oxford University Press, 2018. |
Includes bibliographical references and index.
Identifiers: LCCN 2017059445 | ISBN 9780190246402
Subjects: LCSH: Wetland restoration—United States. | Water—Purification—Biological treatment. |
Sewage—Purification—Biological treatment.
Classification: LCC QH541.5.M3 L44 2018 | DDC 577.68—dc23
LC record available at https://lccn.loc.gov/2017059445

9 8 7 6 5 4 3 2

Printed by Sheridan Books, Inc., United States of America

For Maya, growing strong

CONTENTS

ACKNOWLEDGMENTS

In the course of researching this book, I've been fortunate to spend time with a small army of fascinating people. I can't list them all, but I'd like to offer thanks to Dan Hauser, Wesley Chesbro, Alexandra Stillman (née Fairless), Bob Gearheart, Frank Klopp, Bill Bertain, and the late Dan Ihara: the founders of Humboldt County's wastewater rebellion of the 1970s.

As the book grew more global, I spoke with people farther afield. I'm grateful to Marion Joseph, Phyllis Faber, Bill Mitsch, Grace Brush, and George Tchobanoglous for sharing their knowledge and experience.

Edie Butler and Carly Marino, librarians at Humboldt State University's Humboldt Room, have helped tremendously with locating documents and photographs.

Leslie Scopes Anderson generously shared her time and her beautiful wildlife photographs.

Thanks to Jeremy Lewis, my patient editor at Oxford University Press.

Parts of Chapters 7 and 10 originally appeared in *Undark*. Chapter 11 is based on articles that first appeared in the journal *BioScience*.

Last and always, love and gratitude to Hugh.

INTRODUCTION

The hunter's eye holds a look of obsessed concentration. Ankle-deep in water, the great blue heron waits for its prey. Imitate: Stand still, keep silent, watch. The bird's head shoots forward on its long neck, moving so fast the details of the action cannot be comprehended by human sight. A splash. Now the heron grasps a struggling fish cross-ways in its bill. The victim's tail fin twitches, the fish translates into a silver flash falling into the bird's gullet. A brief digestive pause, then the heron launches into the air, tucking its lanky legs up against its body, folding its neck into a compact S-curve (Fig. I.1).

At the Arcata Marsh and Wildlife Sanctuary a few blocks from downtown Arcata, on California's north coast, it's easy to stumble into scenes like this one. I first explored the place as a wildlife fanatic. River otters played together, gliding across the water, diving without a splash. Thousands of sandpipers burst into a whirl of coordinated flight, startled by the passage of a hunting peregrine falcon. A short-eared owl cruised over the reeds at dusk, listening for the tell-tale splash of a frog moving in the shallows.

This habitat, crowded with birds, hikers, and picnickers, is fueled by sewage. In the 1970s, a handful of activists in this town on the shore of Humboldt Bay fought a years-long political battle with state officials. They rejected the state's plan for an expensive, energy-intensive regional sewage system. In the process, they stumbled on the idea of using wetlands to filter pollutants out of their wastewater. It's a notion so simple and so effective that it has since been adapted at tens of thousands of constructed wetland projects worldwide.

The marsh is a place of paradoxes: a wildlife sanctuary awash in sewage, a habitat that exists because environmentalists fought state officials enforcing the Clean Water Act. It's one of my favorite places to walk, to look at life, to ponder. It's also become a door, a portal into the past and potential future of our relationship to wetlands and to water itself.

We need constructed wetlands like Arcata's because humanity has a penchant for destroying natural ones. The Earth has seen the loss of more than 64 percent of its native wetlands since 1900.[1] In major agricultural states like California, Ohio, and Iowa, the figure is closer to 90 percent. Swamps and marshes have been drained to create farmland and destroy mosquito habitat. They've been paved over with concrete, replaced by shopping malls and suburban neighborhoods, their disappearance considered a victory.

Wetlands develop at the edges between earth and water: at the shallow margins of bays, lakes, ponds, and rivers. Their plants, adapted to soggy conditions, grow with their roots underwater. A swamp is dominated by trees. A marsh grows rushes, sedges, wildflowers. Wetlands capture pollutants, absorb rising waters in times of flooding, stabilize shorelines, act as an important sink for carbon in the era of global warming, and serve as crucial wildlife habitat.

Conservationists first began to advocate for wetland protection when populations of North American waterfowl began to plummet early in the twentieth century. Migratory ducks need wetlands to shelter and feed them during their long journeys to and from their breeding grounds. Human welfare, it turns out, is also tied to the boggy landscapes we had long condemned as wastelands and sources of disease. It just took a long time for us to notice.

While humanity was busy making wetlands disappear—a major undertaking from the mid-nineteenth century onwards—we were adding mass doses of sewage to the waterways that sustain us. The evolution of sewage reveals urban *Homo sapiens* bumbling along in a struggle to evade the stink and disease carried in its own wastes.

Along the way came flashes of brilliance. Observers in Europe and the US began to realize that benign bacteria can break down the pollutants in sewage, and that these microbes can be harnessed to treat our endless flow of fouled water.

The basics of conventional wastewater treatment were worked out at the turn of the twentieth century but would not be widely used until the 1970s, when an army of citizen activists began to fight for clean water. At the same moment, a handful of ecologists began to recognize the ability of wetlands to capture and break down pollutants. This startling revelation hit scientists working in Germany, the Netherlands, North Carolina, Massachusetts, and Florida. The habitats we had forced out of the landscape were revealed as key to controlling the overload of waste and nutrients flowing out of cities, industries, and farm fields.

In 1974, two newly elected members of Arcata's city council began to resist the state's plan for a regional system that would have ringed Humboldt Bay with miles of sewer lines. They didn't know much about wetlands, but they could see that the project as proposed would be the biggest energy consumer in Humboldt County, that it would spend millions of dollars to pump sewage across long distances, that it would lead to urban sprawl. Those were the reasons they started to fight.

Arcata's sewage resisters stumbled onto the low-cost, environmentally friendly concept of wetland treatment, propelled by political pressures. In a way, they were like William Dibdin, the nineteenth-century London bureaucrat. To save the city of London money, he'd set out to argue that its deluge of raw sewage did not harm the Thames River. He ended up showing, through careful observation and experimentation, that managed communities of microbes could cleanse sewage before it reached the river. Those same microbes are at work in every healthy wetland.

Arcata's activists of the 1970s set out to stop a big steel-and-concrete sewage project and ended up building a wetland that has become the city's most beloved

Figure I.1 Great blue heron hunting at the Arcata marsh. Photo by Leslie Scopes Anderson.

park. Tens of thousands of people hike there every year, and most of them have no clue that they're walking through part of a wastewater treatment plant. They're too busy watching the birds to notice. Many long distance migrants find shelter among the reeds. Three hundred thirty-four species of birds have been sighted at the marsh.

Beneath its calm surface, Arcata's marsh holds lessons on the intertwined histories of wetlands and water pollution. More than 150 years after the first sewage treatment techniques were invented, it has become clear that conventional systems won't be enough to keep our lakes and bays healthy, or our drinking water safe. We have transformed whole landscapes to sustain our way of living, fouling vast watersheds in the process. Solving the problem will take many kinds of change. Wetlands, great and small, will be part of the answer.

NOTE

1. Ramsar Convention on Wetlands. "Wetlands: a global disappearing act." Fact Sheet 3.2 (http://www.ramsar.org/sites/default/files/documents/library/factsheet3_global_disappearing_act_0.pdf).

1

Cholera's Frontiers

Sewage as we know it—the everyday miracle of feces disappearing down the toilet, pushed by a never-ending flow of clean water—is a recent invention. The flush toilet itself has been created, and then forgotten, many times down through the ages. But the grand scheme that we all take for granted—an endless supply of clean water piped in and limitless amounts of dirty water piped out—was thought up by Edwin Chadwick (Fig. 1.1), a British lawyer turned public health crusader, in the 1840s.

Back then the cities of the Old World were awash in human waste. Even the most elegant homes had privies that emptied into cesspits, where decades of accumulated filth sat rotting beneath the parlor floor. The poor lived in tenements where dozens of people might have to share one privy. Chadwick supervised a survey of sanitary conditions in English cities that came up with some amazing statistics. In parts of Manchester there was one privy to every 215 people. Some houses had yards covered six inches deep in "human ordure," which the inhabitants crossed by stepping on bricks.

"Sir Henry De La Beche was obliged at Bristol to stand up at the end of alleys and vomit while Dr. Playfair was investigating overflowing privies," Chadwick wrote of one of his colleagues. "Sir Henry was obliged to give it up."[1]

London had sewers, of a sort: They were open ditches that sloped toward the Thames, and were meant to drain stormwater out of the streets. But by Chadwick's day, the gunk from thousands of overflowing cesspits emptied into these sewers, then oozed its way into the river. Parliament's windows on the riverfront had not been opened in years because of the stench. The Chelsea Water Company, which provided drinking water to many Londoners, still had its intake a few feet from the outfall of the Ranelagh sewer.[2] An editorial in *The Spectator* pointed out that city residents paid the water companies "340,000 pounds per annum for a more or less concentrated solution of native guano."

Chadwick's Sanitary Report of 1842 carefully documented the correlation between bad sanitation and high death rates. Today this idea seems painfully obvious, but in the mid-nineteenth century no one understood why filth could kill.

An eccentric Dutch shopkeeper named Anton van Leeuwenhoek had built the first microscope more than a hundred years earlier, and found that his water

Figure 1.1 Sir Edwin Chadwick. Photo from Wikimedia Commons.

cistern, his skin, and the inside of his mouth were all swarming with live creatures, too tiny to see with the naked eye. Microbes were known to exist, but most scientists remained unimpressed. No one had made the connection between van Leeuwenhoek's discovery and infectious diseases. Despite repeated epidemics of cholera, typhoid, and yellow fever, many prominent doctors did not believe that diseases could be contagious at all.

The medical wisdom of the time held that disease was caused by chemical substances from decaying filth. Breathing in these miasmas could cause deadly illness. "All smell is, if it be intense, immediate acute disease," wrote Chadwick, "and eventually we may say that, by depressing the system and rendering it susceptible to the action of other causes, *all* smell is disease."[3]

Chadwick had a vision of how he would deodorize England's cities. It depended on channeling great volumes of clean water into every neighborhood, then flushing it away once it was dirty. This would mean a restructured water supply system.

No one had indoor plumbing in those days. Instead, people paid water companies for limited access to a well or for delivered water. Chadwick foresaw a newfangled water closet in every household.Complex new sewer projects would have to be built. Instead of the old, rectangular ditches that drained London, he planned to build underground pipes, rounded and tapering at the bottom, so that the sewage would flow fast and carry its smelly load off efficiently—into the nearest river.

He recognized his scheme's drawback: It would pollute rivers all over the country. He hoped to solve this by diverting sewage to farm fields, where it could be used as liquid manure. But his first priority was to get the waste out of town as soon as possible, no matter where it went. He believed this would save thousands of lives.

Chadwick's "sanitary idea," that water and sewers should be provided by one centralized authority, run by himself, remained a political football for seven years. The companies that had a longstanding monopoly on urban water supplies opposed him. So did local officials who resented Chadwick moving in on their turf.

In 1848 Chadwick landed the position of Sanitary Commissioner of London. As he campaigned to win the appointment, cholera was sweeping through Europe and into England. People who appeared healthy would suddenly collapse, going cold in the limbs and blue in the face. Many died within hours of becoming ill.

The epidemic reached London in February 1849. Fourteen thousand Londoners would die before the outbreak subsided. Chadwick unknowingly helped kill many of them. He ordered the city's antique sewers to be flushed as never before. Better to pollute the river, he believed, than to leave the stuff in the sewers where it would give off "foetid exhalations."[4]

Cholera bacteria are passed in human waste, and heavy loads were washed into the Thames, the only water source for thousands of people. As the epidemic raged in August 1849, 5,773 cubic yards of filth were being flushed from the sewers into the river each week, under Chadwick's enthusiastic direction. The monthly death rate shot up from 246 in June to 1,952 in July and 4,251 in August, peaking at 6,644 in September.[5]

It was the first of many such public health disasters. No one had found a good way to deal with the resulting sewage, but the idea of household taps and the flush toilet gained in popularity. By the 1880s, American cities like Boston and Newark were sewering up, flushing their untreated waste into the nearest river. In towns downstream, death rates from cholera, typhoid, and other waterborne diseases soared (Fig. 1.2).

Chadwick, an arrogant, prickly character, made many political enemies. He might never have achieved enough power and influence to change London's sewer system if cholera had not made people desperate. Cholera gave Chadwick followers, who would transform the urban landscapes of Europe and America forever. Their passion for better sewers would begin to save lives in the largest, most crowded urban centers, while they intensified the fouling of waters downstream.

Cholera and water pollution are bound together in a complex evolutionary knot; each has helped to create the other. The dread disease that has swept every inhabited continent during the past 200 years, and still sickens millions and kills thousands in the developing world, is a new variation on an ancient bacterial theme. And it flourishes in nutrient-rich water, especially when that water is contaminated with human waste.

Robert Koch, one of the founding fathers of microbiology, first isolated the cholera bacterium in Egypt in 1883. Viewed through a microscope, the tiny culprit he'd extracted from the intestines of cholera patients looked like a comma; Koch called it the comma bacillus, and noted that it could also be found clinging to aquatic plants in polluted waters. Later, other scientists would name the germ

Figure 1.2 "Death's dispensary," cartoon published in 1866, illustrating new evidence that water transmitted cholera infection. Public domain.

Vibrio cholerae. The mystery of cholera wasn't solved with the identification of *Vibrio*, however. New questions kept cropping up. One of the most urgent concerned where *V. cholerae* hid out between its intermittent crime sprees.

Cholera is an ancient illness: An outbreak with symptoms typical of cholera is described in 2,500-year-old Sanskrit writings. The disease is endemic in the Ganges River delta and along the coast of the Bay of Bengal, where outbreaks occur predictably every spring and fall. In 1817, the first episode of pandemic cholera spread from the Ganges delta near Calcutta, India, into the Middle East,

Europe, and East Africa, killing hundreds of thousands. The outbreak subsided for a few years, than flared again in 1829. Over the next two decades, epidemic cholera struck around the world, following the routes of human travel.[6] It reached England in 1831, with sailors who came to port in Sunderland. In the late 1840s, while Chadwick was campaigning to flush more sewage into the Thames, *V. cholerae* reached America. The disease arrived at the port of New York and headed west in the distressed innards of ambitious pioneers.

The second pandemic subsided in 1850, but cholera has swept around the globe five more times since then, gradually evolving new variations. The most recent pandemic, which has spread to South America and Haiti, began in 1961 and continues today, having affected more than 7 million people. It involves a new strain of cholera bacteria that was first identified in 1992.

Between outbreaks, cholera may vanish from human populations for several years or decades at a time. Researchers searched for a hidden reservoir of cholera, a sign that people or their animals could carry the infection without showing overt symptoms. Then in the 1980s, a team led by Rita Colwell of the University of Maryland found *V. cholerae* alive and well in the waters of both the Chesapeake Bay and the Bay of Bengal, clinging to the shells of copepods, miniscule crustaceans that drift with the currents.

The cholera bacterium and its close cousins, also members of the genus *Vibrio*, live in estuaries all around the planet. Many types of *Vibrio* survive in even the cleanest of tidewaters—and most have nothing to do with people, growing only among the plankton, clams, and fish.

The hunt for the wild *Vibrio* took so long because in hungry times—when the waters it lives in are not overloaded with nutrients from sewage or stormwater runoff—*V. cholerae* stops reproducing and goes into a dormant form. The bacteria can live for a long time this way, waiting for the opportune moment, the right combination of temperature, salinity, and nutrient load that will signal the time to wake up and procreate. When microbiologists looked for *Vibrios* using the traditional technique of culturing germs from water samples, these dormant *Vibrios* didn't grow, and so they remained invisible.

Colwell and her colleagues uncovered dormant *Vibrios* using a new tool. They exposed mice to *V. cholerae*, so that the mice produced antibodies that would latch onto cholera bacteria and nothing else. Then they attached a fluorescent molecule to the antibodies, turning them into microscopic signal beacons.[7] When these fluorescent antibodies were added to samples of estuary water, they showed that *V. cholerae* were not only present, but abundant. Most of the germs clung to the shells of planktonic creatures, especially the copepods, which look like a miniaturized cross between a lobster and a beetle. In photomicrographs of these water samples, glowing microbes outline the shapes of the copepods, like constellations in a night sky.

Many disease-causing microbes need people to survive, but *Vibrios* do not. They are everyday citizens of normal estuaries, at home attached to plankton and aquatic plants. The bacteria nourish themselves off the chitin in copepod shells—they have a special enzyme designed to digest it. *V. cholerae* is tightly linked to copepods, so much so that filtering contaminated water through folded sari cloth, which forms a mesh fine enough to capture copepods, removes more than 90 percent of *V. cholerae* cells as well. Five years after Colwell and her colleagues taught this simple technique to women in Matlab, Bangladesh, cholera infection rates there were halved.[8]

When untreated sewage enters the equation, algae, copepods, and *Vibrios* all turn opportunist. Sewage is rich in nutrients, a jolt of liquid fertilizer. Algae absorb it and reproduce like mad. Copepods, which feed on algae, also boom. A polluted estuary is a *Vibrio* paradise, pleasantly brackish and full of abundant food and handy plankton to rest on. When a lot of people are crowded together at the edge of an estuary, without a good sanitation system, their waste fuels an explosion of the *Vibrio* population.

Along the Bengal coast, the disease strikes twice a year, under conditions that can be traced to the abundance of copepods in drinking water. In the spring, low river flows allow plankton-rich bay water to intrude into the ponds used by local people as water sources. In the fall, monsoon rains swell the rivers, flushing nutrients from the land into the delta and bay. Plankton bloom in the fertilized waters, carrying abundant *V. cholerae* across the flooded landscape. Both drought and monsoons are predicted to intensify as the climate continues to warm. Warm-water plankton species are moving toward the poles as ocean temperatures rise. Manmade climate change now threatens to intensify the risk of cholera epidemics.[9]

Yet, of the diverse strains of *V. cholerae* that can be found in the wild, most are not dangerous to humans. To become successful pathogens, strains of *V. cholerae* must acquire a set of genetic traits that allow them to thrive in the human gut, and to spread from one infected victim to another. Among the critical traits are the genes that code for cholera toxin, a protein that wreaks havoc with the salt balance of cells lining the small intestine. The toxin causes cells to pump electrolytes and water into the lumen of the intestine, causing the severe diarrhea and dehydration characteristic of cholera.[10] One result is that the infected person's gut is awash in nutrient-rich, salty fluid, an ideal growth medium for the cholera bacterium.

Pathogenic *V. cholerae* bacteria also produce pili, long filamentous extensions that enable them to attach to the gut lining. The genes that code for pili production act in concert with those that produce cholera toxin. So among cholera researchers, these attachment factors are known as toxin coregulated pili (TCP). The genetic code for cholera toxin comes from a bacteriophage, a virus that infects some strains of *V. cholerae*. TCP act as a receptor for attachment and infection by the bacteriophage. These relationships suggest that the human gut, the only place where pili are expressed, is the site where *V. cholerae* strains acquire their pathogenic traits.

A recent study tracks the evolution of the El Tor strain of *V. cholerae*, the cause of the seventh, ongoing global cholera pandemic. The El Tor strain occupied human guts as early as the 1890s, though at that early stage it did not make people sick. By the 1930s, El Tor was causing isolated outbreaks in which victims suffered symptoms of classic cholera. But the strain was not yet highly contagious, and the outbreaks subsided. It took decades for this line of *V. cholerae* to pick up the traits that made it into a lethal threat. The first outbreaks of the seventh pandemic occurred in 1961 near the town of Makassar in Sulawesi, Indonesia. The responsible strain had originated decades earlier in the coastal waters of Bengal, and it acquired more and more pathogenic traits as it traveled in human bodies along an ancient migratory pathway that led first to the Middle East and then to Indonesia.[11]

In 2010, the El Tor strain made the leap across oceans to Haiti, where it triggered an intense outbreak of cholera. "The disease struck with explosive force," said reporter Richard Knox. "Within two days of the first cases, a hospital 60 miles away was admitting a new cholera patient every 3-1/2 minutes." Cholera has since killed thousands in Haiti, and is now endemic there as it is in its ancient homeland of Bangladesh.[12] Genomic sequencing of the *V. cholerae* strain involved showed that the disease had traveled to the island in the guts of Nepalese peacekeepers sent by the UN to aid in Haiti's reconstruction following a major earthquake, and had been released into streams and rivers because of improper sewage disposal at the UN camp near the town of Mirebalais.[13]

Throughout its history, cholera has flared up from polluted waters to travel across oceans and continents inside human hosts. In 1867 and again in 1891, cholera struck a large gathering of people who came to the shores of the Bay of Bengal for the Hardwar Fair, and many of the survivors carried the germ back to their hometowns. Several times in the mid-nineteenth century, cholera swept through the thousands of Muslims gathered for the annual pilgrimage to Mecca, and again the people carried pathogenic *Vibrios* back to their homes.

America's pilgrimage in those times was west to the California Gold Rush. People from all over the US dropped everything and followed the immigrant trail. Among them was an opinionated doctor named Israel Shipman Pelton Lord. The journal he wrote during his travels is very much a tale of cholera. Lord lived in Illinois; he'd only gotten as far west as Missouri when, in May 1849, he began to meet victims of the disease.

"We left a dead man by the name of Middleton on the levee at St. Louis, and thought that we had left all the cholera with him," Lord wrote. "We were grievously disappointed, however."[14] Along the Missouri River, he watched some of his fellow steamboat passengers die off, and passed towns that had been emptied by the scourge. Later, as he moved through Kansas and Nebraska, he found cholera victims lying exhausted and alone at the trail's edge.

By September 1850, Lord had tried his hand at gold mining in California and given it up in disgust. He was living in Sacramento, a hectic, filthy boomtown near the spot where gold had first been discovered (Fig. 1.3). The previous spring, the town had been home to 150 people. The Sacramento Lord knew that fall held 6,000, and what primitive sanitation there may have been could not keep pace with the population explosion. When the Sacramento River flooded during the boom years—and it often did—it spread waterborne diseases everywhere. Eventually, the settlers would raise the street level by twelve feet in an effort to escape their own sewage.

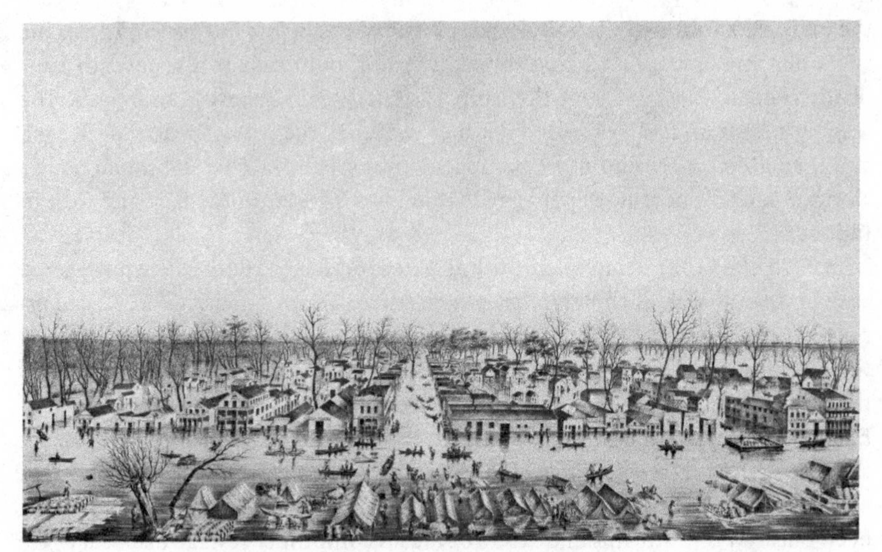

Figure 1.3 View of Sacramento, California, during a flood in January 1850. Floods intensified the risk of waterborne disease in Gold Rush boom towns. Drawing by George Casilear and Henry Bainbridge. Courtesy California History Collection, California State Library.

"The cholera is approaching us slowly from both east and west," Lord wrote on September 30, 1849. "It will make a charnel house of Sacramento when it comes." He was soon run ragged trying to attend to cholera victims, and recorded as many as forty deaths in a single day. His diary entries included long, angry harangues against doctors who dosed cholera patients with laudanum, which he believed did more harm than good.

On November 3, his only diary entry was terse: "Sick today. Cholera symptoms. Weather the same." Lord survived. By late November, when the disease had run its course in Sacramento, he estimated that three-fourths of the city's population had died.

In December 1849, Israel Lord suffered through a flood on the polluted Sacramento River. The stink of the sewage-enriched Thames choked London. And Josiah Gregg stood knee-deep in a pristine stream just north of Humboldt Bay. He was cursing like a lunatic. In a few years this river, named the Mad in honor of Gregg's fit of temper, would form Arcata's northern boundary, and farms would cover the lowlands along its banks.

Gregg led a band of eight ragged men who came west from their gold mines on the Trinity River, staggered for weeks over snow-covered mountains and through a maze of giant redwoods, and arrived on the shore of the Pacific just north of Arcata. They were starving, cold, and extremely irritable, and they were about to go down in history as the first white men to discover Humboldt Bay by land.

An adventurer and scholar who had written a popular book about his travels on the Great Plains, Gregg had an obsession for scientific measurement. This, along with the severe trials of the journey, was driving the men with him crazy.

While they were recovering from some nervous encounter with a band of Indians who had never met white people before, or fainting with hunger after two days without food, Gregg would insist on stopping to measure a redwood or to read their latitude.

By the time they reached the Mad River, they had no patience left for Gregg. When he stopped to read the latitude at the mouth of the river, they cursed him soundly—Lewis K. Wood, who survived the expedition to become one of Arcata's founding citizens, remembered that this was not at all unusual—and threatened to launch their canoes without him, leaving Gregg stranded. He had to wade into the water to catch the canoes, and on the other bank he exploded in rage.

Gregg had spent most of his life following the frontier west, beyond the civilized boundaries of the US. "I could never live under my oppression of spirits anywhere in the U.S. where I would be liable to continued annoyance," he once wrote to his brother. But he ended his life by opening up the last untamed corner of California to settlement.

He died trying to make his way back out of the redwood wilderness, and was buried near Clear Lake, about halfway between Humboldt Bay and San Francisco. His companions claimed that starvation killed him, but some historians speculate that they murdered him out of sheer aggravation.

Either way, Wood and the other members of the expedition did make it back to San Francisco, and they brought word of what they'd found: a bay surrounded by fertile lowlands, a nice spot where a man could farm, if he was not too busy getting rich by running supplies to the gold mines on the Klamath and Trinity rivers. Civilization came to Humboldt Bay in a rush.

By 1850, Arcata's town square had sprung up on a small prairie overlooking the bay, where the Gregg expedition had spent Christmas Eve in 1849. Within a few years, stores and saloons lined the edge of the plaza, and more than 500 people lived in town.

The scale of the place was still small enough that people could get along fine with an outhouse in the backyard. Nobody had indoor plumbing. When folks needed water, they carried buckets past the cows grazing in the middle of town and pumped at the well that had been sunk at the center of the plaza.

No one worried about piping in water until 1875, when a fire in Alexander Brizard's store raged out of control and destroyed most of the businesses on the plaza. The town board of trustees decided Arcata needed water mains, the first ingredient for a functioning fire department. After years of dithering, the water system was finally completed in 1884. Pipes, made of hollowed-out redwood logs laid end to end, carried water from holding tanks on the hill above town down to four fire hydrants on the plaza.[15]

The water supply wasn't just used for the fire hydrants, of course. The town's wealthier families were building elegant homes for themselves, and they wanted running water on tap. They also began to order water closets, ornate porcelain fixtures that came up from San Francisco by ship. The flush toilet hit the big time in the 1880s. By the end of the decade, few people would consider building a house without one.

In Arcata, as in cities all over the US, people started to build sewers as an afterthought, when they realized they needed somewhere for the stuff coming out of their toilets to go. The town's first sewer line ran down from the big Victorian houses on the hill, past the new brick store that was Alexander Brizard's pride. It went down to the edge of the bay, where the raw sewage came gurgling out.

NOTES

1. Finer, S.E. (1970). "The life and times of Sir Edwin Chadwick," p. 2 and 34. *New York: Barnes & Noble, Inc.; London: Methuen & Co Ltd.*
2. Halliday, S. (1999). "The Great Stink of London: Sir Joseph Bazalgette and the cleansing of the Victorian metropolis," p. 21. *Stroud, England: Sutton Publishing.*
3. Johnson, S. (2006). "The ghost map," p. 114. *Riverhead Books, New York.*
4. Chadwick, E. (1842). "Report on the on the sanitary condition of the labouring population and on the means of its improvement" http://www.deltaomega.org/documents/ChadwickClassic.pdf.
5. Bingham, P., N.Q. Verlander, M.J. Cheal (2004). "John Snow, William Farr and the 1849 outbreak of cholera that affected London: a reworking of the data highlights the importance of the water supply." *Journal of the Royal Institute of Public Health* **118**: 387–394.
6. Drasar, B., B.D. Forrest (1996). "Cholera and the ecology of *Vibrio cholerae*." *London, New York: Chapman & Hall.*
7. Colwell, R. (1996). "Global climate and infectious disease: the cholera paradigm." *Science* **274**: 2025–2031.
8. Colwell, R., et al. (2003). "Reduction of cholera in Bangladeshi villages by simple filtration." *PNAS* **100**: 1051–1055.
9. Levy, S. (2015). "Warming trend: how climate shapes *Vibrio* ecology." *Environmental Health Perspectives* **123**(4): A82–A89.
10. Levy, S. (2005). "Cholera's life aquatic." *BioScience* **55**(9): 728–732.
11. Hu, D., Bin Liu, Lu Geng, et al. (November 14, 2016). "Origins of the current seventh cholera pandemic." *PNAS*, E7730–E7739.
12. Domonoske, C. (2016). "U.N. admits role in Haiti cholera outbreak that has killed thousands." *NPR* (http://www.npr.org/sections/thetwo-way/2016/08/18/490468640/u-n-admits-role-in-haiti-cholera-outbreak-that-has-killed-thousands).
13. Orata, F., Paul Keim, Yan Boucher (2014). "The 2010 cholera outbreak in Haiti: how science solved a controversy." *PLoS Pathogens* **10**(4): e1003967.
14. Lord, I.S.P. (1995). "At the extremity of civilization: an Illinois physician's journey to California in 1849." *Jefferson, North Carolina:McFarland & Company.*
15. Trainor, J. (1984). "The first hundred years: Arcata Volunteer Fire Department, 1884-1984." Arcata, California: Arcata Volunteer Fire Department

2

Tides of Change

When Dan Hauser and his friend Wesley Chesbro won the Arcata city council race, their opponents did not concede gracefully. "I'm not a poor loser," claimed Clyde Johnson, just before he called Hauser and Chesbro "rangatangs." Then Johnson and the other disappointed candidates accused the winners of using dirty campaign tricks—just like President Nixon. Arcata's weekly paper, the *Union*, ran the details of the post-election flap on its front page.

That March of 1974, the national obsession with the Watergate scandal reached its peak. The president's closest aides were on trial for burglary, wiretapping, and obstruction of justice. Nixon had become an international symbol of corruption, and the polls showed his public approval rating plummeting to an all-time low. So while Hauser and Chesbro could laugh off the comparison to an ape, when they were likened to the president the insult cut deep.

It was a rough time to start a political career, especially in Arcata, an old logging town on the shores of Humboldt Bay in California's damp northwest corner. The community was splitting in two like a redwood slat struck with an ax. On one side stood ranchers and timber workers, many of them descendants of the first pioneers to settle here in the 1850s. On the other were outsiders like Hauser and Chesbro, people who'd recently migrated to town to study or teach at Humboldt State University (HSU), and who'd decided to stay in this foggy enclave, 250 miles north of San Francisco. Now, for the first time, the outsiders controlled the city council.

The old-time Arcatans felt like victims of an alien invasion. That feeling intensified when the national fad for high-speed nudity reached HSU. A few days after the election, four young guys ran naked through the University quad. Behind them, the crowns of the redwood trees at the edge of campus vanished into the fog. A cold rain fell as the earnest exhibitionists moved across the lawn, and goosebumps rose all over their bodies. They were members of the Streakers for Impeachment Committee, a national organization dedicated to streaking for the ouster of President Nixon.

Dan Hauser was a mild-mannered insurance adjuster and liberal Democrat, caught in the clash between the buck-naked protesters on campus and traditional Arcata families. The accusations of "Watergate-style dirty tricks" against him and

Chesbro were wildly overblown. The losers were reacting to a skit broadcast on the university radio station just before the election. It mocked all the contenders but was kind to Hauser and Chesbro. The broadcast, the defeated candidates complained, "crossed the lines of fair play and good taste."

Hauser had known nothing about the radio sketch until after it was aired. He had spent the campaign walking all over town, knocking on doors and explaining his stand on land-use planning to anyone who would listen. On days when he couldn't summon up the will to go out and campaign, his eleven-year-old daughter gave him a pep talk and went along with him.

Two months before the 1974 election the Brizard-Matthews Machinery Company, owned by two long-time Humboldt County families, bought a full-page ad in the *Union*. They used the space for a political call to arms. Their ad accused environmentalists of wrecking the local economy by fighting to put more forests in national parks, and of raising food prices by working to "ban DDT and other necessary pesticides." It urged readers to get out and vote—against people like Hauser and Chesbro.

The two new council members disagreed with Arcata's conservatives on everything from the Vietnam War to forestry reform. They won the election with the help of a strong voter turnout from HSU students, who'd never wielded much influence in town before. But after a change in federal law granted eighteen-year-olds the right to vote in 1972, Arcata politics would never be the same.

The old Arcata stalwarts knew it, too. The night before the council election, Hauser and Chesbro walked up to campus to post their campaign literature. Everywhere they went they found the walls already covered with flyers that urged the students not to vote. *This isn't your community*, the flyers read. *You shouldn't determine the city's government.*

The "Don't Vote" propaganda was Lois Arkley's idea. Lois, the daughter of Robert Matthews of Brizard-Matthews, worked the campus that evening with her husband Robin, a conservative mill owner and rancher. The Arkleys stood united in their loathing of the political changes they saw coming. "Chesbro was born for us to hate," recalls Robin Arkley. "Here was this hippie kid with long hair and a scraggly beard, who was going to be running our town. We felt we were being totally disenfranchised."

"My first reaction was to tear those posters down," says Hauser. "But Wesley, being more politically astute than I was, said we should leave them up, because it would make the students mad enough to get out and vote. And in the end it did have that reaction. So we followed along behind the Arkleys. And we put our posters up next to theirs."

Hauser is a tall, soft-spoken man, a shy person whose stomach used to clench when he had to speak in public. It took a lot of Rolaids to get him through the city council campaign. In 1974, he'd been married for ten years to his high school sweetheart, Donna, and they had two young kids.

When he was growing up in southern California, Hauser watched urban sprawl devour the farmland around his home in Orange County. By 1961, when he started college, the truck gardens and orange groves he'd loved as a kid were wiped out. “Traffic was becoming impossible,” he says, “and it was not where I wanted to raise a family.” He discovered Humboldt County during a fishing trip with his dad, and two years later, he and his new wife moved north.

His campaign for city council focused on resisting uncontrolled development. Arcata was still surrounded by second-growth forests and pastures dotted with dairy cows. Seven miles of farm fields stretched along the edge of the bay between Arcata and Eureka, the nearest city to the south, and Hauser wanted to keep them open.

If you pulled off of Highway 101 and spent a few minutes in those fields, you’d see much more than cows. A red-shouldered hawk would pass overhead. Then a great egret, three feet tall and draped in startling white plumes, would stalk by. The big birds liked to loiter in pastures, where they found easy prey. Field mice gave themselves away when they ran from the crushing hooves of the cattle.

Hauser knew that developers were eager to change this bucolic scene. The city of Eureka had recently tried to annex a large chunk of farmland, hoping to pave the way for a series of fast-food places, motels, and shopping centers. The tide that had washed away rural Orange County waited to sweep into Humboldt.

He had campaigned against a big development that the Fords, an old-time Humboldt County ranching and logging family, planned to build south of Arcata. “It was the wrong development in the wrong place,” he says. “I’d done my homework. The bay muds would not support the kinds of structures they wanted to build at that location.” There was no sewer system there, either, so the waste from the apartment complex and shopping center the Fords envisioned would have nowhere to go. Ultimately, Hauser’s efforts defeated the development, but they also earned him some dedicated enemies.

A group of long-time Arcatans formed an organization they called Arcata Forever, and proceeded to spend the next few years making life difficult for Hauser and Chesbro. “They saw us as somewhere very far to the left of Josef Stalin,” says Hauser. “We were anti-Vietnam War, so we were therefore anti-patriotic.”

By the time of his first city council meeting, Hauser had seen enough to know that his critics would be loud and uninhibited, but he thought he knew what to expect. “My issues were land-use planning and historic preservation,” he says. “I knew some of that would be controversial but I thought the majority of the people would support it.”

He knew nothing about the regional sewage project that appeared on the agenda at that first meeting. He didn’t know that he’d spend the next six years fighting against the big sewage plant, while every authority figure in sight assured him his cause was hopeless.

“We’d had no preparation for this issue, and Wesley and I balked, things were going over our heads,” says Hauser. “At the end of the meeting, a friend came up to me and just blasted me for not taking a strong stand. I said, ‘I don’t even know what they’re talking about.’ ”

In all the technical documents that had been thrown at him, one thing caught Hauser’s eye. It was a diagram that showed a sewer line running from Arcata south to Eureka along Highway 101, at the edge of the pastureland he’d been trying to protect. If that sewer line was ever built, the cow pastures would be turned into a developer's field of dreams. Anyone who paid attention knew that if you built sewers, they'd come.

Humboldt Bay was in trouble: That much was undeniable. A 1973 survey found sixty-three different pipes oozing mysterious fluids into the water, sources unknown. The city of Eureka had three separate sewage treatment plants, none of which worked reliably or reached modern standards. In the hard rains of winter, treatment capacity was overloaded and raw wastewater flowed into the bay. The creeks rose, carrying loads of silt and manure downstream from the dairy pastures.

Humboldt County was in step with the times. All over the US, aquatic creatures choked on human waste. Fish in the Great Lakes floated belly-up by the thousands. The surface of the Cuyahoga River in Cleveland, smothered in oil from industrial discharges, burst into flame.

Nothing so dramatic happened on Humboldt Bay. Instead it was oysters that raised the alarm, like silent canaries in a subaquatic coal mine. In January 1973, a few months after Congress had passed the Clean Water Act, state inspectors found that oysters in the bay were contaminated with fecal coliform bacteria, microbes that thrive in human and animal feces. The entire bay was quarantined for months.

Humboldt Bay had become California's oyster capital after the once-thriving San Francisco fishery died out in the early 1900s. The native oysters there were gone, suffocated in a big-city deluge of raw sewage. Subjected to only a fraction of San Francisco Bay's human population, Humboldt's waters were relatively clean. The oyster farming industry got rolling in the 1950s, when some local entrepreneurs began to ship seed oysters in from Japan. The shellfish thrived in the fertile waters off Arcata and Eureka. Oyster farmers boating out to their underwater fields became an everyday part of life on the bay. Business boomed until the state inspectors shut it down that winter of 1973.

Oysters spend their entire adult lives anchored to one spot, and eat by pumping water through their guts, filtering out and digesting bits of algae, bacteria, and debris. Any microbe in the water will accumulate inside oysters. By the 1960s, scientists had begun to use them as living barometers of sewage pollution. Oysters were no longer just a popular delicacy on the half shell; they offered undeniable proof that Humboldt Bay, like so many other bodies of water, was tainted.

On a gray afternoon in the spring of 1973, Robert Rasmussen stood at the doorway of the Ingomar Club, an ornate Victorian mansion so close to Humboldt Bay that he could hear the tide sloshing against a cement seawall. Bearded and rotund, Rasmussen resembled a youthful Santa Claus, but he was feeling tense rather than jolly. He loitered, examining the complicated curlicues of the building's wooden trim. He'd never been this close to it before, didn't belong in this bastion of the county's most powerful men.

Rasmussen, a botany professor moonlighting as an environmental consultant, had only been invited to lunch at the Ingomar because, as he explains it, "The shit was hitting the proverbial fan."

Waiting inside the Ingomar was a small knot of men in suits. The group centered around David Joseph, the director of the Regional Water Quality Control Board, a short, stocky man exuding an air of command. He'd helped to write clean water policies for the entire state of California, and he believed there was only one way to address the problems in Humboldt.

Joseph wanted a high-tech, centralized treatment plant that would bypass the bay and discharge straight into the Pacific. All the sewage from Arcata and Eureka would have to be routed to one site, piped under the floor of the bay, treated and released into the ocean. There'd be lots of expensive design and construction work involved. The engineers from Metcalf & Eddy, who sat with Joseph amid the polished oak and crystal of the club, were glad such a plum job had dropped into their laps.

The men waiting for Rasmussen were expecting him to act as one more cog in the machinery the Clean Water Act of 1972 had set in motion. The legislation provided grant money to states for the construction of sewage treatment facilities—a lot of grant money. Confident that the feds would pick up much of the tab, Joseph and his colleagues had written a Basin Plan for the north coast, an edict that forbade any discharge of sewage into the bay, treated or not. The cities of Eureka and Arcata had clubbed together to hire Metcalf & Eddy to design a new system for central Humboldt County, so that they could comply with the Basin Plan. Metcalf & Eddy hired Bob Rasmussen's little firm, Environmental Research Consultants, to write the Environmental Impact Report, a new-fangled document required by the recently passed California Environmental Quality Act. Rasmussen was the smallest fish in a vast bureaucratic pond.

A heavy door swung open on well-oiled hinges, and Rasmussen was ushered into the Ingomar's high-ceilinged dining room. He sat across from Ron Robie, an influential young water attorney with California's State Water Resources Control Board. Years later, Rasmussen, having forgotten Robie's name, would inaccurately describe him as "Joseph's sycophantic junior."

"That whippersnapper told me this would be the easiest money I'd ever make," says Rasmussen. "All I had to do was follow the guidelines they'd already written, agree that a regional system was the only way to go." Rasmussen nodded at Robie and made vague, affirmative noises. Then he went out and wrote a report that said the state strategy was all wrong.

Humboldt Bay, he argued, had always been a natural sump. Every winter's rains washed nutrient-rich silt off the hills, into the creeks, and down to the bay. The muddy bay bottom grew a thick crop of eelgrass, the wide green blades swaying in unison, bending in the direction of the prevailing tide. The eelgrass sheltered an array of mollusks, worms, and young fish, and fed thousands of waterfowl that passed through on their long migrations. The plants, rooted in a deep black goo composed of their own decaying leaves, of dead saltmarsh plants and algae and small water creatures, formed an important habitat, one worthy of protection. It was also, Rasmussen wrote, a habitat "adapted to defecation and decay."[1]

The bay, he argued, was better suited to absorb treated sewage than the ocean. As long as the treatment process removed harmful bacteria, a bit of effluent should do no harm. Existing studies suggested that sewage did not constitute a significant addition to the rich stew of natural nutrients in Humboldt Bay. By contrast, sewage discharges to the open ocean, which were already standard procedure on the southern California coast, had been shown to alter the ecology of the waters near outfall pipes. Algae bloomed; the number of marine species dwindled.

Rasmussen's report was a long riff of criticism against the conventional approach to sewage pollution. Whether effluent was discharged to the bay or the ocean, dumping the stuff rather than recycling it would prove to be a temporary fix, not a long-term solution, he insisted. The idea that the temporary fix had to be ocean disposal was based on politics, not biology.

Despite the sixty-three anonymous discharge pipes, the plants in Eureka leaking raw sewage, and the steady flow of dissolved cow manure, the only immediate problem in the bay was bacterial contamination. Yes, the communities at the water's fringe needed to clean up their act, treat their sewage and disinfect it. But there was no reason, other than Joseph's edict, that they had to build a pipeline to carry their effluent beyond the bay.

The engineers at Metcalf & Eddy were stunned when they read Rasmussen's report. He took the skeptical engineers out onto the bay and showed them the abundant growth of eelgrass rising out of the muck. He convinced them.

"The benefits of ocean discharge are illusory," the Metcalf & Eddy staff wrote in the documents they submitted to the cities of Arcata and Eureka. They recommended that Eureka build one high-capacity treatment plant that could handle all the city's sewage. Arcata's plant, which was already in good working condition, needed only a minor upgrade to ensure that the system did not overload during the heaviest rains. Both facilities, sited on the bay's edge, could release treated effluent right into its murky waters. This scenario made the most sense in terms of both economy and ecology. The only potential obstacle was David Joseph and his crusade to forbid any wastewater at all from entering the bay.

The Metcalf & Eddy report, with its subversive recommendation buried among pages of technical analysis, was presented to the Arcata city council in March 1974, at the first meeting after Dan Hauser and Wesley Chesbro were elected. Months passed while the complexities of the sewer problem percolated in their brains.

They had plenty of other problems to distract them. In April, Hauser, Chesbro, and Rudi Becking, an HSU forestry professor who also served on the city council, passed a resolution calling for an end to US involvement in the Vietnam War. The members of the conservative contingent in town were outraged and never missed a chance to let the three responsible council members know it. Becking was a World War II veteran, survivor of a brutal Japanese POW camp. Well known as a straight talker with a powerful tendency to piss people off, he seemed impervious to criticism. Hauser and Chesbro, however, felt the sting of some very personal attacks. One night, after a meeting ended, Robin Arkley—the mill owner who'd papered the campus with posters admonishing students not to vote in the city council election—drew the two of them aside. In the midst of a heated conversation, Arkley flipped open his billfold, revealing a snapshot of Chesbro. "I carry this around," he announced, "so I can remember who to hate."

Chesbro was juggling a complicated mixture of loyalties. He'd filed for conscientious objector status and was doing alternative service at the Northcoast Environmental Center in Arcata. His environmentalist friends had serious doubts when he and Hauser began to talk about resisting the state's sewage plan. The Clean Water Act may not have been perfect, but for activists struggling to clean up polluted waters, it was the best opportunity to come along in years. Some influential environmentalists in town worried that if the city council fought the state, they'd lose that chance.

In early September, after a city council meeting that had gone on past midnight, Chesbro, Hauser, and Alexandra Fairless, Arcata's first woman mayor (Fig. 2.1), shared a few beers at the Jambalaya. The Jam was a dim hangout just off the plaza, a two-block walk from City Hall. It was the bar favored by college students and others among the younger, longer-haired residents of town. When he wasn't doing research for the Environmental Center, Chesbro worked as the Jambalaya's bouncer.

The three of them gathered around a rickety table, exhausted and wired. The meeting at City Hall had been full of intense debate. "Council meetings in those days went on for a long time because of Alex," remembers Hauser. "She enjoyed engaging people in argument and baiting her opponents."

Arcata's City Manager had just informed the council that fighting the state sewage plan would be a mistake. The city would have to bring a legal challenge against Joseph's Basin Plan, and odds were high they'd lose. If they didn't jump on the state bandwagon now, they risked losing any shot at the federal money designated for sewage improvements under the Clean Water Act.

Fairless wasn't strong on the technical details. When Hauser joined the council, she was happy to pass on to him the responsibility to represent Arcata in negotiations over the proposed regional sewage system. But she understood well enough that Joseph's grand vision of centralized treatment was flawed. The country was in the midst of an energy crisis; long lines formed to get a few gallons of fuel at the local gas stations. A single sewage plant for the Humboldt Bay area would be the biggest energy consumer in the county. The pumps needed to move wastewater over the miles between Arcata and Eureka, and then across the bay to the open ocean, would suck down more power than the busiest lumber mill.

In the late-night dusk of the Jambalaya, the three friends hashed the issue over for the hundredth time. The state refused to consider moving the proposed sewage pipeline in order to prevent development on the pastures that Hauser had worked so hard to preserve as open land. Joseph's plan looked to be a destructive waste of money and energy. Rasmussen, the resident expert, said all this was unnecessary; they could protect the bay in humbler, less damaging ways.

It's hard to say exactly when the city council majority decided to fight despite the odds. Maybe the resolve came to each of them separately, on the quiet walk home from the Jam, or in that peaceful instant before they opened their eyes and rolled out of bed the following morning.

Figure 2.1 Wesley Chesbro, Alexandra Fairless, and Dan Hauser in an Arcata parade early in their tenure on the city council, 1974 or 1975. Photo courtesy of the Alexandra Fairless Campaign Collection, Humboldt State University Library.

What is on the record is that at the next week's meeting, the council voted to direct the city attorney to determine whether Arcata would stand any chance at all in a legal battle with the state. They hired Rasmussen and his colleagues to try to poke more holes in David Joseph's arguments, hoping to sink them completely.

Becking, fearless as usual, sided with the council's young radicals. "If the council has any guts at all," he said, "we should challenge the state."

Part of the reason the city council resented the state's costly regional sewage plan was that Arcata had already sunk serious money into upgrading its treatment plant. In its fifty acres of newly built oxidation ponds, microbes broke down organic matter while winds off the bay helped to keep the water oxygenated. The ponds attracted hundreds of ducks—Arcata's treatment plant was the most popular birding spot in the area—and served the same function as the conventional, high-tech activated sludge treatment systems preferred by the regional board. The city was carrying over $1 million in bond debt incurred in building the oxidation ponds and installing a chlorination system to disinfect effluent. If forced to join the regional system, Arcata would have to pay millions for its construction and maintenance while continuing to pay off its investment in a local treatment plant rendered useless.

Instead, Arcata hoped to improve its existing plant, while finding a way to recycle the treated wastewater. A fisheries professor from HSU had been experimenting with raising young salmon and steelhead in treated sewage effluent. The city council hoped the intense use of its oxidation ponds by migrating birds, along with the recycling of effluent in aquaculture, could justify an exemption from the regional project.

Fairless asked state officials to meet with Arcata's city council to discuss alternatives to the regional sewage system. David Joseph attended, along with Robie.

"We hoped we could develop more advanced biological treatment here in town," remembers Fairless. "David Joseph asked what my degree was in, and I told him I was a Home Ec major." Decades later, the memory of Joseph's response still makes her bristle. "When you get your degree in biology," he said, "I'll be glad to discuss this issue with you."

Joseph and his engineer at the regional board, John Hannum, had both helped shape statewide regulations. They were passionate supporters of a rule called the Bays and Estuaries policy, enacted in May 1974, which banned all discharges of sewage to California bays unless the effluent could be shown to "enhance" the waters. (San Francisco Bay, though badly polluted, was exempted from the rule—officially because of its "high tidal exchange" and "the depths of its waters." However, those familiar with the political workings of the time suggest the true reason was that a discharge ban was not remotely possible in such a large, urbanized estuary.)

"Humboldt Bay does not need anything your waste plant can discharge," Joseph told the city council, "including distilled water."[2] If Arcata found a way to meet state water quality standards without joining the regional system, he pledged to have the standards changed to force them into line.

Hauser explained his struggle to keep open space between the cities of Arcata and Eureka. Running a sewage line through the miles of pasture land between the two cities would guarantee the kind of urban sprawl he'd uprooted his family to escape. "The Emperor Joseph sat and listened to my passionate plea," Hauser says. "And he said, 'You can't win.' It was clear that going against him would mean a big fight."

Joseph had grown up on the southern California coast. He witnessed the devastation of marine life choked by pollution from sprawling cities. He had a passion for righting that wrong. His strategy, however, did not take Humboldt's local conditions into account.

The small cities scattered at Humboldt Bay's edge were home to only 40,000 people. Timber and fisheries (and to some extent, the new industry of illegal backwoods marijuana farming) drove the local economy. The bay was alive, hosting rays, halibut, harbor seals, and a commercial oyster industry that provided two-thirds of the oysters harvested in California.

Bob Rasmussen, the marine botanist who'd been among the first to oppose the regional plant, wrote a scathing critique of the state's plan. He pointed out that recent studies by the state Department of Public Health found no significant bacterial contamination from local sewage plants. Runoff from cattle pastures was the major source of the coliform bacteria sometimes found in bay oysters—a problem that the regional sewage system would do nothing to address. He questioned the supposed economies of a large-scale project. The same kind of facility had been dismissed as far too expensive in the Metcalf & Eddy report only a few months earlier. Arthur Einerfield, an engineer with Metcalf & Eddy, had said that locating a regional treatment plant on the Samoa Peninsula, as the regional board recommended, would court disaster by pumping raw sewage under the bay floor. A pipe failure could make the dire predictions of bay pollution a reality in minutes.

The authors of the Basin Plan claimed that without an ocean discharge, Humboldt Bay waters would soon become oxygen-depleted, making the bay unlivable for fish. They forecast that the bay would experience a deadly syndrome, known to aquatic ecologists as eutrophication. A heavy load of nutrients from sewage would cause a dense algal bloom; masses of algae would die off and sink to the bottom, where the bacteria decomposing their remains would use up all the oxygen in the water. The result would be an underwater dead zone where no aquatic creatures could survive. Eutrophication had been well documented in Chesapeake Bay, Long Island Sound, and other polluted estuaries on the East Coast. But in forecasting the fate of Humboldt Bay, the plan quoted data that seemed to be lifted from a report on the much more urbanized and polluted San Francisco Bay. Very little information on the water quality or ecology of Humboldt Bay existed, and what there was came from studies by Rasmussen and his HSU colleagues. They'd found communities of bottom-dwelling creatures, plankton, and marine plants consistent with a healthy ecosystem.

The bay, Rasmussen argued, was using the nutrients in sewage, which served as a replacement for a vast acreage of vanished salt marsh that once would have enriched the waters with decomposing plants and the feces of hundreds of thousands of waterfowl. The same effluent that nurtured the bay would act as a destructive shock to marine life in the open ocean off the Samoa Peninsula, which was not adapted to high nutrient loads. Milt Boyd, a marine ecologist at HSU who testified against the proposed regional sewage system, agreed.

Looking back, Boyd acknowledges that the HSU scientists of the 1970s had very little evidence for their position. Years later they would learn that the dominant cordgrass in surviving Humboldt marshes was in truth an exotic species, introduced in ballast water from ships that sailed between South America and Arcata in the late 1800s. Later studies would show that healthy salt marsh does act as a nutrient sink during spring and summer, when native plants are growing. Plants die back in the fall, and winter rains wash their remains into the bay where they become a nutrient source.

People on both sides of the argument were describing Humboldt Bay in terms of their own fears and visions. "The 70s was an era when you could say almost anything, because our knowledge base on estuaries was so thin," says Jim Cloern, a US Geological Survey expert who has been studying the shifting ecology of San Francisco Bay since 1976. In 2001, Cloern published an article on the evolving science of eutrophication in estuaries that debunked the notion that all bays could be protected in the same way.[3] He compared conditions in San Francisco Bay to those in Chesapeake Bay, which is far more impaired by eutrophication even though it carries lower concentrations of nutrients. A number of factors made San Francisco Bay more resilient, among them booming populations of introduced clams and mussels that filter algal cells out of the water, and strong tidal flushing that carries oxygenated water down to the bay's floor.

At the time of Humboldt County's wastewater battle, however, most scientists assumed that all bays and estuaries were pretty much the same, that there was a simple relationship between nutrient load and the syndrome of algal blooms leading to hypoxic waters. Rasmussen and his colleagues argued early on that Humboldt Bay experienced "tremendous flushing action" with the changing tides, which would move nutrient-enriched waters into the ocean and recharge the bay with oxygenated water. Studies by HSU oceanographer John Pequegnat had produced evidence to support that notion, but Joseph found it easy to dismiss.

Joseph had been struggling to get Humboldt Bay communities to clean up their act for years. In 1966, early in his tenure as executive director of the regional board, Joseph issued a cease-and-desist order to push the city of Arcata to install a chlorinator to disinfect effluent at its treatment plant. In 1970, Joseph cracked down on three Eureka property owners, including the Humboldt County Planning Office, for discharge of raw sewage into the northern arm of Humboldt Bay. The regional board had given repeated warnings before issuing a cease-and-desist order. When County Supervisor Don Peterson asked why cease-and-desist orders were being issued when sewer line construction was already under way, Joseph gave a blunt answer: "We don't trust you."

By the mid-1970s, the regional board estimated that 3.5 million gallons of inadequately treated wastewater were dumped into Humboldt Bay daily. Joseph had spent decades jousting with companies and cities that seemed willing to do anything to avoid investing money in cleaning up their pollution. To him, Arcata's arguments against the regional system represented more of the same. Joseph remembered vividly a spill of raw sewage that had burst from a broken pipe in Arcata into the bay in the winter of 1973. Arcata's public works director had found the leak quickly but failed to notify the regional board, or local oyster farmers, of the contamination.

Joseph and his colleagues hewed to the letter of the Bays & Estuaries policy they'd helped to establish. Sewage discharges to California's bays, no matter how well treated, were forbidden unless they could be shown to enhance bay waters. Asked for a definition of "enhancement," Joseph replied, "The ultimate enhancement would be no discharge at all." A few weeks later Joseph provided a definition of enhancement in formal bureaucratese: "A demonstration that the discharge, through the creation of new beneficial uses or a fuller realization, enhances water quality for those beneficial uses which could be made of the receiving water in the absence of all point source waste discharges."

City council members believed that, given time, they could prove enhancement. The regional board, however, refused to give them time. Designing an alternative treatment scheme and proving it would enhance the bay could take years, but the city had only ninety days. If Arcata didn't join with the regional system by

January 1, 1975, it faced a ban on all new building in the city and fines of up to $6,000 per day.

For a few weeks, the Arcata city council teetered on the brink of rebellion, despite dire warnings from their attorney of the extreme financial consequences of resisting the regional board. In late October, the Arcata *Union* ran a long piece explaining the sewage controversy, along with an editor's note urging readers to absorb it all in order to understand "perhaps the most important [and costly] question ever to face Arcata."

The array of pumps needed to move sewage for miles across flat land and the activated sludge treatment plant itself would have combined to form by far the biggest energy-consuming enterprise in the county. Today, looking back at old newspaper clippings on the issue, it's ironic to see adamant statements by the regional board's representatives on the need for a centralized sewage system quoted next to a photograph of cars lined up to buy gas during the oil crisis of 1973–74, when the oil embargo by the Organization of Petroleum Exporting Countries (OPEC) inspired in US citizens a sudden interest in energy conservation.

The stress roiled Hauser's innards. He made regular trips to the Fourth Street Market in Arcata to stock up on antacids. "Leo down at the Fourth Street Market used to tell me I should buy stock in whoever manufactured Rolaids," he remembers.

In November, the city reluctantly agreed to comply with the regional board's policy. "I'm convinced the discharge requirements were set just so we couldn't meet them," Hauser told a reporter.

When the Humboldt Bay Wastewater Authority (HBWA) was created in January 1975, Hauser was designated Arcata's representative. The authority hired Winzler and Kelly, a local engineering firm, to draw up a plan for a regional system. At the start, controversy erupted between Arcata and the other HBWA members, which included Eureka, McKinleyville, and Humboldt County. The state had recommended a sewer line running from Arcata along Highway 101 to Eureka. Hauser and McKinleyville's representative, Grant Ramey, wanted an alternative route that would carry sewage from their towns through an interceptor on the Samoa Peninsula directly to the treatment plant, which was to be built on the peninsula near the Simpson Pulp Mill. This scenario would avoid development in the pastures along Highway 101 and save Arcata and McKinleyville hundreds of thousands of dollars. But the HBWA board voted 4–2 for the original state design—motivated in part by the fact that it forced Arcata to help pay for an expensive underwater pipeline that would carry Eureka's waste across the bay to Samoa. See Figure 8.2.

"I don't think [the other HBWA members] meant to," said Ramey, "but they were setting Arcata up, because they had them hooking in and helping Eureka pay for their costs and then pumping it under the bay. The whole system might have run if they had not come up with this. But this infuriated Arcata."[4]

The longer the planning process took, the higher the price tag climbed. When the final design was approved in December 1975, the estimated cost was $21 million. In the fall of 1976, grants from the federal Environmental Protection Agency

(EPA) and the state came through, enough to cover $38.5 million in construction costs. By then, however, estimated costs had more than doubled to $52 million, due to inflation. The HBWA board approved the issuance of $12 million in local construction bonds. By then, opponents were pronouncing the project's name as *HUB-wah*, in ominous tones that suggested some fearsome, mythical beast.

HBWA was impelled by the powerful forces of bureaucratic inertia and the longstanding assumptions of profit-driven engineers. Much as Arcata's rebels loathed the idea, the construction of the regional sewage plant appeared inevitable.

NOTES

1. Anderson, R., Robert Rasmussen, Conrad Recksiek, et al. (1974). "Eureka-Arcata regional sewage facility project." Environmental Impact Report Environmental Research Consultants, Inc., Arcata, California (Humboldt Room, Humboldt State University Library).
2. Brisso, P. (1974). "Arcata not needed for wastewater funds to be granted." Arcata Union October 2, 1974.
3. Cloern, J. (2001). "Our evolving conceptual model of the coastal eutrophication problem." *Marine Ecology Progress Series* **210**: 223–253.
4. Scott, E., Leslie Brunetta. (1989). "Wastewater Wars." *Kennedy School of Goverment Case Program* **C16-89-854.0**.

3

The Microbe Solution

In the hot, dry summer of 1858, the Thames was a stew of sewage that festered in the sun, giving off an unbearable stench. "We believe this to be the uncleanest, foulest river in the known world," wrote a London pundit in July. "There you shall see in the brief space of half an hour and two or three miles, a hundred sewers disgorging solid filth, a hundred broad acres of unnatural, slimy chymical compost . . . The water—the liquid rather—is inky black."[1]

Dockworkers suffered nausea, headache, sore throats, temporary blindness—some of them fainted from breathing in the river's aroma.[2] In the newly rebuilt Houses of Parliament, on the riverbank, legislators choked on what the press labeled "the Great Stink." The Thames had been badly polluted for decades, but the heat and low water that summer brought the situation to a crisis. Benjamin Disraeli, leader of the House, held a handkerchief over his nose as he fled from the Chamber, complaining that the Thames had become a "Stygian Pool."[3] In July 1858, he introduced a law that authorized the construction of a costly new sewer system, designed by engineer Joseph Bazalgette, that would carry London's waste downstream of the city.

Britain's rivers were overwhelmed with sewage, its cities bursting at the seams. Between 1801 and 1841 London's population had grown from 958,000 to 1,948,000. Numbers of people living in smaller cities like Leeds, Bradford, and Huddersfield doubled or tripled in the same span of time. While the same pattern held in other European and American cities, geography made the problem more intense in Britain, where the rivers were too small to carry off the wastes of the towns that sprouted on their banks. In 1885, engineer James Gordon estimated that dumping the raw sewage of the major towns along the Rhine would give that river a concentration of only one part sewage per 2,345 parts water. The lower Lea, a tributary of the Thames whose upstream flows had been diverted to provide drinking water for London, was by contrast composed of two-thirds sewage.[4]

The extreme pollution took a heavy toll on aquatic life. An overload of nutrients in raw sewage fueled algal blooms that created oxygen-depleted dead zones, suffocating fish. The last Thames salmon was caught in 1833. By 1890, nine thousand river basins in England and Wales were devoid of fish. The River Almond in Scotland was inundated with waste from shale oil works, and flames thirty feet

Figure 3.1 Monster Soup: A woman drops her teacup in horror after viewing the microscopic beasts in a drop of polluted Thames River water. Drawing by William Heath, 1828. Courtesy of Wellcome Library.

high were reported rising from its waters—a phenomenon that would be echoed decades later on the burning surface of the Cuyahoga River in Ohio.

Sewage was a pressing problem in nineteenth-century Britain, at a time when the idea that microbes could both cause disease and digest sewage was not understood. Politics dominated science, a pattern that has been repeated during struggles to protect water quality ever since. Experts were recruited to argue for interested parties in court cases and at government hearings, often shading their testimony to fit the cause of their clients. Progress toward solutions to the crisis was also impaired by widespread beliefs that now seem quaint, but at the time inspired intense argument over matters that truly meant life or death for both rivers and the people who lived along them.

One such idea was the theory of zymotic disease, which held that decay was catching. In dead matter, some experts argued, contact with oxygen or anything already rotting caused complex organic molecules to begin to shake themselves apart. Putrefaction could cause disease if the vital force in a living creature was weaker than the disturbance caused by contact with rot. "The essence of the concept of zymotic disease," notes the historian Christopher Hamlin, "was that disease was a spreading internal rot, that it came from an external rot, and that it could be transferred to others."[5]

The influential chemist Justus von Liebig was a strong advocate of the zymotic disease theory, but he also understood that decay was an inevitable part of natural cycles. Plants needed nutrients from rotted organic matter to grow—he was among the first to understand that crops require minimum amounts of specific elements, including phosphorus and potassium.

Edwin Chadwick, the famous cheerleader for water closets and modernized sewers, believed that decay was an evil that should be kept away from humanity at all costs. The only place for sewage or anything else liable to rot was in the soil; there human waste could feed crops, but only if it reached the farm in a fresh condition. In Chadwick's view, sewage needed to flow quickly away from houses and towns. Disease arose from the smelly gases generated by organic matter putrefying in stagnant sewers. At an 1870 meeting of the Institution of Civil Engineers, Chadwick described, in approving tones, the sight of fresh feces and toilet paper flowing into the Thames from newly built, self-cleansing tubular sewers at Croydon. In his eyes this was a vast improvement on the homogenized brown muck that oozed from the antique sewers in London, where waste sat in stagnant pools until a rainstorm washed it down to the river. Chadwick was certain that Croydon's unrotted waste would fertilize the river, just as it could fertilize farmland if it were piped there. "Whilst old sewage that was putrid killed fish, sewage that was fresh fed them," he declared. At Carlisle, after new tubular sewers were installed, Chadwick claimed that "the fish were not only increased in quantity by greatly improved in quality. Anglers now found their best sport at the mouths of the tubular outfalls discharging sewage fresh."[6]

The notion that decay was an inevitable part of nature's cycles disturbed many who believed in natural theology, which held that a benevolent Creator had designed the world for humanity's use. This philosophy led its adherents to tie themselves in some intriguing intellectual knots.[7]

In the midst of London's Great Stink, William Odling, a chemist and health officer to the Vestry of St. Mary, Lambeth, argued that the expense of the new London sewer system proposed by engineer Joseph Bazalgette was not needed, because microscopic animalcules in the water would devour the organic matter in sewage, rendering the Thames clean. These tiny creatures would magically prevent the gunk fouling the river from putrefying, and cause it to vanish.[8]

Like other believers in natural theology, Odling refused to let the foul reality of the Thames change his conviction. It would take the combined impact of the two major advances in biological science of the late nineteenth century—Darwin's theory of natural selection, along with the discovery that microbes could cause disease—to loosen the chokehold of natural theology on many Victorian minds. (As a young man, Darwin himself had accepted that all things in Nature were designed by a benevolent Creator.)[9]

Because decay seemed contrary to this divine order, British sewage experts would long strive to prevent the decomposition of sewage—a hopeless ambition that prevailed in the years before the basic ecology of bacteria was understood. When oxygen is abundant, aerobic bacteria decompose organic matter into harmless, odorless components: carbon dioxide, nitrates, and water. This is the process at work in every modern sewage treatment plant, as it is in every healthy river, bay, or wetland. In the absence of oxygen, anaerobic bacteria dominate—and while they too break down organic matter, they do so through a different chemical process that releases the nasty-smelling gas hydrogen sulfide, the source of the rotten-egg stink of polluted waters. The Victorians called this odiferous process "putrefaction." They were both surrounded and horrified by it.

Parliament approved Bazalgette's plan, and construction of his sewers was completed in 1860. The new system was an impressive feat of engineering that relocated London's sewage downstream to the Thames estuary.[10] The Thames also continued to receive sewage from towns upstream of London. Throughout Britain, downstream cities took to suing their upstream neighbors over the fouling of water supplies. London's Metropolitan Board of Works, dumper of the city's sewage, collected evidence of pollution by others in the Thames estuary, in case of a lawsuit.[11] Eventually, "Bazalgettianism" became an insult applied to those who advocated dumping sewage without treating or recycling it.[12]

Chadwick, the original advocate for more efficient sewers in London, had always envisioned city sewage being used to irrigate and fertilize farmland. The idea made intuitive sense; before the coming of the water closet, London had a longstanding tradition of scavengers who made their living by carrying night soil to surrounding farms to be used as manure. The same was true of many cities in India, China, and parts of Europe.[13] Dumping sewage into rivers or the sea meant an irretrievable loss of valuable fertilizer.

Liebig, the chemist, was also a determined advocate of using sewage as manure on croplands. In the introduction to his book, *Agricultural Chemistry*, Liebig wrote that the rise of water closets in England led to the annual loss of nutrients capable of producing food for 3.5 million people, while the expensive guano farmers imported from South America as fertilizer ran off the fields and into rivers, adding to the pollution problem. In his eyes, the flush toilet had created a parasitic nation. "England," he wrote, "like a vampire it hangs upon the breast of Europe, and even the world; sucking its life-blood."[14]

Enthusiasts like Chadwick and Liebig believed that sewage was such a valuable fertilizer that it could be sold for great amounts of cash—enough to pay the enormous expense of sewer construction. Chadwick tried to prove the worth of the idea in 1849, when he formed the Metropolitan Sewage Manure Company, which planned to use some of London's sewage for irrigation. The company never turned a profit and folded a few years later, having lost £50,000 in capital.[15]

Filtering sewage through the soil did a fair job of cleaning it up, but it was a nuisance for farmers. The sewage came day after day, an unending flow of tainted water, whose volume increased during wet winters when it was least needed on the land. In many cases an excess of sewage flooded the fields, cutting off the supply of oxygen to bacteria in the soil and creating a stinking landscape of anaerobic decay. Even when they worked, sewage farms required vast expanses of suburban land. Experts advised at least one acre of farmland per 100 people in the city generating the sewage. Under this formula, 1850s London would have needed 40,000 acres of sewage farm adjacent to the city, where real estate was at a premium. (Some cities did have successful sewage farms, most notably Berlin, Germany, where farmworkers drank effluent flowing off the fields. Such farms functioned as treatment systems first and croplands second. Proof of the Berlin farm's effectiveness as a treatment system came during a typhoid epidemic: While the waterborne disease ravaged the city, workers on the sewage farm stayed healthy.)[16]

Fed up with the failure of many sewage farms to prevent water pollution, Britons turned to a different set of experts who claimed they could use chemistry to spin a profit from wastewater. The idea was that the right chemical recipe would cause organic matter to settle out of sewage, producing a convenient, dry fertilizer while leaving a clean effluent that would flow to rivers without causing a nuisance. Four hundred eighty patents for processes related to extracting nutrients from town sewage were taken out in Britain from 1850 to 1890.[17] Most used substances such as lime, charcoal, phosphate rock, clay, and alum to precipitate organic matter.

These processes failed. Heavy loads of organic pollutants remained dissolved in the effluent. Worse, the strong chemicals used killed the aerobic bacteria that decompose organic matter. Victorian chemists who believed that decay was evil may have been pleased with this state of affairs, but it was only temporary. Once the precipitated sewage flowed into a river, the harsh chemicals were diluted and the process of decay kicked in again. Much of the organic matter in the sewage remained, fertilizing blooms of algae. Soon the waters were drained of oxygen, so that anaerobic bacteria took on the work of decomposition, emitting the stink of putrefaction.

Bazalgette's sewers carried waste out of London and released it, untreated, at two outfalls on opposite banks of the Thames, at Barking and Crossness. The sewage flowed up and downstream with the tides. At low ebb, banks of stinking black mud emerged from the water. People who lived and worked along the lower Thames complained often of the stench and the dead state of the river, but nothing changed. In 1878, the *Princess Alice*, a cruise ship, sank near the outfalls. Locals claimed that the lost passengers died not of drowning, but of immersion in a river composed mostly of sewage.[18]

The *Princess Alice* disaster, along with a petition signed by 13,000 lower Thames residents, triggered a series of hearings held by the Royal Commission on Metropolitan Sewage Discharge, created to consider the impacts of dumping London's raw sewage into the Thames estuary. In May 1882, as the Metropolitan Board of Works was bracing to defend itself at the upcoming hearings, a young chemist named William Dibdin was appointed to head the Board's chemistry department. As an argument for continued sewage discharge, Dibdin resurrected the old notion that minute creatures in the river would get rid of the pollution. At first this was a political dodge to evade the great expense of intensive chemical treatment, or of creating a large sewage farm on the lower Thames. After the Royal Commission issued its verdict, finding that London's sewage was indeed fouling the estuary and would have to be cleaned up, Dibdin scrambled to find an affordable solution. In the process, his understanding of microbes and water pollution evolved.

His work was inspired by two witnesses who testified before the Royal Commission. Henry Clifton Sorby, an amateur microscopist, had studied tiny crustaceans, called entomostraceae, and used their numbers as an index of pollution in the Thames.[19] Sorby found that fecal particles from entomostraceae were abundant in polluted stretches of the river, and deduced that the creatures fed on human waste. He proved this point by experimentation, keeping entomostraceae in a tank with human excrement as the only food source. "I have kept a number of them for six weeks," he told the Royal Commission, "and you could see that they are healthy and happy and in good spirits; you could see them eating human excrement all the time."[20] Sorby believed this was strong evidence that the creatures helped to remove sewage pollution. (Fig. 3.1).

The respected chemist August Dupre, testifying as a consultant to the Metropolitan Board of Works, suggested that pollutants in the Thames would be removed through the action of even smaller life forms: aerobic bacteria. Dupre was among the first to achieve this insight, and later collaborated with Dibdin in creating England's first biological sewage treatment filters.

In the years following the Commission hearings, Dibdin studied oxygen concentrations and the corresponding presence or absence of sewage and of fish at different points along the Thames. He experimented with sewage treatment at the Barking outfall, and discovered that filtering sewage through heaps of piled rock removed most of the pollutants. Dibdin created a one-acre bed of coke breeze that, if managed well, could successfully treat 4.5 million gallons of sewage per week. (Coke breeze—chunks of partially burned coal—happened to be abundant and cheap; coarse gravel would have worked just as well. The filter served as a habitat for aerobic bacteria, which formed films on the surfaces of the coke and did the work of decomposing the waste dissolved in sewage.) By 1892, Dibdin had come to understand the microbes in sewage, or in the river, as a living community with distinct needs for nutrients and oxygen:

> By reason of the multitude of organisms to be dealt with, a filter may be fittingly compared to a great animal. If you overfeed it, the result will be disastrous. If you overfeed the river with effluent, then you put more organic matter into the river than the aquatic life is capable of dealing with. The food on which the aquatic life lives, if in excess becomes corrupt . . .[21]

These ideas, which remain the basis of modern sewage treatment, were established only after great political and intellectual struggle. As early as 1860, Louis Pasteur had shown that microbes are responsible for fermentation, the spoiling of broth or milk. Nobody generalized this concept to infer the role of bacteria in breaking down organic matter in soil and water until 1877, when two French chemists, Theophile Schloesing and Achille Muntz, proved that nitrification—the oxidation of organic matter to carbon dioxide, water, and nitrates—was a transformation performed by microbes.[22]

Schloesing had studied sewage farms outside Paris, and wondered what mysterious property of soil purified the water. He and Muntz designed a simple experiment. They filtered sewage through a glass container filled with sterile sand and limestone. Nitrates appeared in effluent from the container only after twenty days.

Sterilization of the sand with chloroform vapors halted the nitrification, which started up again only after the sand was inoculated with washings from fresh soil. The researchers concluded that nitrification was performed by microbes; the twenty-day delay occurred because it took that long for a thriving population of bacteria to become established.

Nitrification, which takes place only in aerobic conditions, had been regarded as the ultimate process of purification since the 1840s. The findings by Schloesing and Muntz, which were soon repeated by British researchers, seemed to some to vindicate the belief that nature was designed for the convenience of humanity. The Earth would be cluttered with corpses, fallen leaves, and kitchen scraps, in addition to an infinite amount of undecomposed excrement, had the Creator not had the foresight to stock the planet with these benevolent microbes.

Still, at the moment in history when the germ theory of disease was becoming widely understood, the notion of bacteria cleansing sewage did not reassure everyone. Some feared that nitrifying bacteria might also be pathogens, while others felt sure these useful microbes must also devour disease-causing germs at the same time they broke down waste. In an 1893 editorial, *The Lancet* opined that the right kinds of bacteria would power the sewage treatment plants of the future; still, for the present, it was best to try to obliterate them all.[23]

Dibdin predicted that bacteria would be harnessed to treat sewage in industrialized systems.[24] In a paper he read to the Institution of Civil Engineering in January 1887, he wrote, "when this subject is better understood, in all probability the true way of purifying sewage . . . will be to turn into the effluent a charge of the proper organism, whatever that may be, specially cultivated for the purpose . . ." Some attendees of the meeting laughed out loud, and Dibdin was thoroughly mocked for his statement. If he was wrong about some details, he proved right about the gist: Modern sewage treatment plants are carefully managed habitats for aerobic bacteria.

Dibdin's research at the Barking outfall was a painful process of trial, error, and argument. At one point, a colleague assigned to run the one-acre filter bed overloaded it with sewage, causing the whole system to clog and putrefy. Dibdin and his assistant, George Thudichum, let the bed rest for three months, then slowly filtered increasing volumes of sewage through it, allowing time for a thriving population of bacteria to become established on the surfaces of the chunks of coke. They monitored levels of oxygen in the effluent to make sure they didn't load too much sewage too fast. They achieved a filtration rate of more than 1 million gallons of sewage per day through the one-acre bed, with 80 percent of the pollutants removed.[25] "The whole mass of the filter remains perfectly sweet and free from putrescent organic matter," reported Thudichum. "The life of the filter appears to be without limit, provided always that the balance between aeration and food supply to the organisms is preserved."

The same principles were worked out across the Atlantic, at the Lawrence Experiment Station, established in 1886 by the Massachusetts State Board of Health. The Station's mission was to find the best way to treat sewage in order to rescue the state's increasingly foul waters. Boston Harbor had received the city's raw sewage for decades, and was far too contaminated to swim in. Similar conditions prevailed in many of the older cities in the US. A newspaper editor described the odor of Baltimore's inner harbor in the hot summer of 1897: "It is a 2000-horse-power smell that lays limburger cheese in the shade."[26]

London chemists and sanitarians were limited by politics and a chronic lack of funding for research. At Lawrence, scientists and sanitarians had the freedom to take a more systematic approach, testing the effects of different types of filtration media from sand to coarse gravel, comparing bacteria found in their sewage filters to those found in nature, and studying the environmental conditions that favored bacterial growth.

William Sedgewick, a professor at the Massachusetts Institute of Technology, and his former student Edwin Jordan led the biological studies.[27] Both were Darwinists, and applied the concept of natural selection to the ecosystems in their sewage filters. The microbes living in filters came from water and soil in the surrounding environment, but Jordan expected that not all species found in nature would thrive there:

> The sewage itself—a nutritive medium of varying composition and richness—will contain only those species capable of living and holding their own in the continual struggle for existence . . . The chemical composition of the sewage undoubtedly debars some species from taking part in the contest.[28]

Jordan knew the germs that inhabited the sewage filters at Lawrence in intimate detail. It was his job to coax them into visibility. He did this by mixing a sample of sewage effluent with nutrient gelatin, pouring the mixture onto a sterile glass plate, and waiting for bacterial colonies to grow. Each colony represented the offspring of a single parent germ. Under a microscope, he could see the shape of individual cells. Cocci were spherical, bacilli rod-shaped, spiral bacteria twisted. The color of the colony offered a clue to the nature of the bacterium—*Bacillus coli communis*, one of the most common microbes in sewage, made red colonies. (Now known as *Escherichia coli*, this bacterium was known to be a normal resident of the human colon, as well as the guts of other animals, and is still used as an indicator of sewage contamination.) Jordan took small samples from a single colony and studied a microbe's response to different growth media: milk, potato, agar. Could the microbe grow without air? Did cells sprout flagellae and swim? What temperatures could it tolerate? All of these factors had to be worked out before he could label the bacterium as a member of a particular species.

In the early 1890s, Jordan took a position as an instructor at the University of Chicago. By the time he was named Professor of Bacteriology in 1907, he'd spent years immersed in the city's complex sewage problems.

Chicago's population had mushroomed from about 350 in 1833 to over 1 million in 1890.[29] Raw sewage generated by the growing population, along with the blood and guts discarded by the city's stockyards, ran into the two branches of the Chicago River, which emptied into Lake Michigan. A branch of the river's South

Fork, downstream of the meatpacking district, was known to locals as Bubbly Creek. Upton Sinclair described it:

> One long arm of the river is blind, and the filth stays there forever and a day. The grease and chemicals that are poured into it undergo all sorts of strange transformations, which are the cause of its name; it is constantly in motion, as if huge fish were feeding in it, or great leviathans disporting themselves in its depths. Bubbles of carbonic acid gas will rise to the surface and burst, and make rings two or three feet wide. Here and there the grease and filth have caked solid, and the creek looks like a bed of lava; chickens walk about on it, feeding, and many times an unwary stranger has started to stroll across, and vanished temporarily.[30]

The city had been built on flat marshland, and the river's flow was so sluggish that winds off the lake sometimes pushed the polluted water back upstream into town. This changed during times of heavy rain, however, when pollution from the river flowed into Lake Michigan, the source of drinking water. "In the great rain of August 2, 1885," noted a report on Chicago's sewage, "the contents of both branches of the Chicago River, with the sewage accumulation of many weeks . . . were, in a few hours, belched incontinently into the lake."[31]

The city's water department had built a tunnel that extended two miles into Lake Michigan, designed to pull clean drinking water back to shore. Pollution from the river often spread far into the lake, however—far enough to contaminate water drawn into the offshore tunnel.

Chicago had high numbers of typhoid fever deaths compared to other major cities in the US and Europe. Typhoid fever is caused by the bacterium *Salmonella typhi*, which, like the cholera germ, thrives in sewage-contaminated water. Typhoid attacks the gut, causing vomiting, diarrhea, dehydration, and high fevers. In severe cases, the disease can cause coma and death.[32] In 1891, Chicago lost 1,997 citizens to the disease, and about 20,000 people suffered infection—a situation experts described as "an epidemic of really alarming proportions."[33]

Chicago had just been chosen to host the Columbian Exposition, an event expected to draw hundreds of thousands of visitors. Jordan's mentor William Sedgewick, along with Allen Hazen, a young chemist trained at the Lawrence Experiment Station, studied the history and causes of typhoid epidemics in the city, pointing out the danger of disaster if visitors were exposed to Chicago's tainted drinking water. They cited evidence that the city's makeshift sewage system failed on a regular basis. Chicago used pumps, located at Bridgeport, to lift the sewage-laden waters of the Chicago River into the Illinois & Michigan Canal, built in the 1820s to allow shipping between the Great Lakes and the Mississippi. This strategy had been discovered by accident, when workers pumping water out of the river to raise the level of the canal noticed they'd reversed the river's flow. The pumps pushed Chicago's filth over the natural divide between watersheds, sending it into the Des Plaines River, which flowed into the Illinois, which emptied into the Mississippi.

"The cause of the enormous excess of typhoid in May 1891," wrote Sedgewick and Hazen, "was the total stoppage of the Bridgeport pumps . . . allowing sewage to flow directly into Lake Michigan."

Forewarned of the risks of drinking Chicago tap water, the organizers of the Columbian Exposition built their own water purification plant, as well as importing clean water from a famous spa in Waukesha, Wisconsin. In an inspired public relations move, they hired Hazen to set up a lab that tested the quality of the Exposition's drinking water daily. Hazen's data showed the Exposition's water was much safer than the city's. In May 1893, for example, city tap water contained 630 bacteria per cubic centimeter, while Waukesha water had only 204, and the figures for filtered water were even lower.[34]

A few years later, city officials decided to permanently reverse the Chicago River's flow by cutting a new canal that would carry its foul waters into the Illinois Valley. The channel would be large enough to accommodate shipping traffic, replacing the old Illinois & Michigan Canal. Its construction involved a mammoth excavation. One booster described it as "a mighty channel which will rank with the most stupendous works of modern times."[35]

The city of St. Louis, Missouri, which lies just below the junction of the Illinois and Mississippi rivers, had also been suffering typhoid epidemics. In an effort to stop Chicago from flushing its sewage into the watershed that provided drinking water for St. Louis and other cities on the Mississippi, the state of Missouri filed suit in what became the first pollution case decided by the US Supreme Court.[36] In classic Chicago style, the Sanitary and Shipping Canal was opened by stealthy crews of men armed with shovels, who broke down the dams holding back their city's filthy waters before dawn on January 17, 1900, the day St. Louis' attorneys were to file for a court injunction to stop them. The Chicago River began to flow backward into the Des Plaines, diluted by relatively clean water from Lake Michigan. A *New York Times* report on the events bore the headline, "The Water in the Chicago River Now Resembles Liquid."[37]

Jordan, who'd begun his career by studying the workings of biological treatment in intricate detail, testified in support of Chicago's right to re-engineer its river and let its raw sewage flow down to the Mississippi. He took samples from points scattered along the hundreds of miles of river between Chicago and St. Louis. The typhoid germ was difficult to isolate and identify, so Jordan used the number of *E. coli* colonies cultured from each sample as an indicator of sewage pollution, and as a stand-in for the typhoid bacterium.[38] "The best index for gauging the continued vitality of the typhoid bacillus in running water is information from the fate of the colon bacillus," he explained.

Jordan (Fig. 3.2) found abundant *E. coli* in the Sanitary Canal at Chicago, and as far downstream as the town of Morris on the Illinois River. The water also held high concentrations of organic matter in the early stages of decomposition—the stuff that fed the booming population of *E. coli*. It carried high concentrations of ammonia, a volatile form of nitrogen typical of heavily polluted waters. But at Ottawa, twenty-four miles downstream of Morris, the *E. coli* had almost vanished. Through the action of aerobic bacteria, the ammonia had been transformed into nitrate.[39] This improvement was due in part to dilution from the Kankakee River, which joined the Illinois just above Morris, and partly to the river's natural ability to cleanse itself through the action of aerobic bacteria.

By the time the Illinois flowed past the town of Averyville, signs of the intense pollution from Chicago had disappeared. The water quality was comparable to that of relatively pristine tributary streams, like the Kankakee, Fox, Vermilion, and Sangamon rivers. The Illinois suffered another dose of filth when it passed through the city of Peoria, with its stockyards and distilleries. Yet by the time it joined the Mississippi, at the site of the town of Grafton, the river had shrugged off the new load of pollution and had lower bacterial counts than the main stem of the Mississippi. The self-purification of rivers, Jordan testified, "is not an interesting biological myth, but an actual and definite occurrence."

The great majority of bacteria in the Sanitary Canal at Chicago did not survive long enough to reach Ottawa. Based on Jordan's findings, it was a fair assumption that any increase in typhoid bacteria that occurred in the Mississippi upstream of St. Louis was far more likely to be caused by pollution from Peoria. What counted in dealing with sewage pollution, Jordan said, was the destruction of pathogenic bacteria, not levels of organic matter or chemicals in the water. By this standard, the Illinois River had shown a remarkable ability to cleanse itself.

Figure 3.2 Edwin O. Jordan, the pioneering bacteriologist who studied the ability of rivers to cleanse themselves of bacteria from sewage. Photo from Wikipedia.

In 1906, Justice Oliver Wendell Holmes wrote the Supreme Court decision in the case of *State of Missouri v. State of Illinois and the Sanitary District of Chicago*. The question at hand, Holmes said, was "whether the destiny of the great rivers is to be the sewers of the cities along their banks."[40] The court's answer to this question was a resounding "yes." Chicago had the right to pollute the Illinois River, even though the city had dug an artificial canal to bring its sewage into the watershed. American cities spewed raw sewage into rivers throughout the nation. The burden of proof rested not with Chicago, the polluter, but with St. Louis, the plaintiff. In a long battle of experts, St. Louis had failed to prove that Chicago's waste caused significant harm to people drinking tap water drawn from the Mississippi.

In the aftermath of the court decision, St. Louis built a water purification plant, using a design developed at the Lawrence Experiment Station. Mississippi River water passed through beds of sand that filtered out bacteria, providing safe tap water. The right to pollute rivers became the legal and political norm in the US; if a city wanted safe drinking water, it would have to build itself a filtration plant as St. Louis had done.

There were objectors. In 1910, the Pennsylvania State Board of Health required Pittsburgh to build new sewers that would separate household wastewater from street runoff, along with a treatment plant. The city hired Allen Hazen and another well-known sanitary expert, George Whipple, to make a plan of action. Hazen and Whipple's influential report concluded that replacing Pittsburgh's sewers and building a treatment plant would cost at least $46 million. The twenty-six towns downstream from Pittsburgh on the Ohio River could provide filtered water for their residents for much less. "No radical change in the method of sewerage or of sewage disposal as now practiced by the City of Pittsburgh is now necessary or desirable," they wrote.[41] The Board of Health backed down, and the status quo of raw sewage discharge continued. Communities that did bother with sewage treatment went only as far as primary treatment:, building simple tanks where the solids would settle out of standing sewage. Few imitated the efficient sewage filtration beds that had been created at Barking and Lawrence.

Chicago's population quickly ballooned into the millions, and the impact of the increasing sewage load could be seen and smelled on the Illinois River. In the summer of 1911, two biologists from the Illinois Natural History Survey observed "septic conditions for twenty-six miles of the course of the Illinois from its origin . . . the water was grayish and sloppy, with foul privy odors distinguishable in hot weather. Putrescent masses of soft, grayish or blackish slimy matter, loosely held together by threads of fungi, were floating down the stream."[42] They might have been describing many other rivers and estuaries downstream of major cities in the US.

The same year, officials in New York were scrambling to remedy intense pollution in the city's harbor. They invited Gilbert Fowler, a chemist and bacteriologist from Manchester University in England, to consult on the problem. Fowler had been studying a curious microbe his colleague Ernest Mumford had discovered in polluted Manchester canals. Given air, iron, and a source of nitrogen, the bacterium, dubbed M7, broke down proteins and created deposits of iron compounds. Fowler and Mumford had hopes that M7 could be harnessed to treat sewage.

After his visit to New York, Fowler traveled to Lawrence, which he described as the Mecca for sewage researchers. Harry Clark, the chief chemist at Lawrence, had been trying to determine how much sewage fish could tolerate and still survive. His team had found that as they increased pollution levels, fish would die unless increasing amounts of air were bubbled into the aquaria.[43] That observation had led them to abandon their studies of fish and move to studies of the effect of aeration on sewage itself. By bubbling air into bottles of raw sewage, they could purify it in only twenty-four hours.

Back in Manchester, Fowler asked his assistants, Edward Ardern and William Lockett, to repeat the sewage experiments he'd seen in Massachusetts. Ardern and Lockett put raw sewage in a bottle and bubbled air into it until all the nitrogen had been converted to nitrate, the stable form that was the mark of an effluent that would not putrefy when released into a river. The process took six weeks, far too long to be useful in treating city sewage. But Ardern and Lockett went a step farther: They poured off the purified sewage, saving the sludge that had accumulated on the bottom of the bottle. Then they added a new batch of raw sewage, bubbled air into it, and poured off the purified liquid again, saving the gunk at the bottom. After repeating this process several times, they'd created a powerful brew that, in the presence of plenty of oxygen, could purify raw sewage in about six hours. The process needed no contact beds, which require substantial space—a problem in cities where every scrap of land is valuable.

Ardern and Lockett had created a living stew of aerobic microbes capable of rapidly breaking down organic matter and oxidizing nitrogen; they called it "activated sludge." The power of the sludge was in the organisms it contained. The researchers proved this by showing that sterilized batches had no ability to cleanse sewage. When Ardern presented their findings at a 1914 meeting of the Society of Chemical Industry, his colleagues saw it as an "epoch-making" advance in sewage treatment.[44]

Soon after, Robbins Russel of the University of Illinois isolated the nitrifying bacteria in activated sludge. They proved to be familiar organisms, first identified in 1890 by the great Russian microbiologist Sergei Winogradsky. Winogradsky had shown that the decomposition of organic matter in soil involves a chain of bacteria, each performing a different chemical reaction. Among them were microbes that gathered their life force not from harvesting the energy of sunlight, as green plants do, nor from feeding on organic matter, as every creature from protozoans to people do. These bacteria harvested energy from the chemical reactions of inorganic substances, in a process called chemosynthesis. The first such group of microbes he isolated were what Winogradsky called his "beautiful oval bacteria."[45] They were *Nitrosomonas*, the genus that metabolizes ammonia into nitrite in soils, and in sewage, around the world. Later he isolated *Nitrobacter*, the germ that oxidizes nitrite into stable nitrate.

Nitrogen, like carbon, oxygen, phosphorus, and sulfur, is essential to life. It is everywhere—more than 70 percent of Earth's atmosphere is composed of molecular nitrogen gas, N_2. Yet nitrogen is among the least available of the primary nutrients. In the elemental gas, two nitrogen atoms are connected with powerful chemical bonds, which can be broken only by a bolt of lightning—or by a few kinds of nitrogen-fixing microbes. These bacteria convert N_2 gas to ammonia, NH_3, a form that can be absorbed by plants. Once this happens, nitrogen enters the realm of living things, where a single N atom may cycle over and over through many different forms. It may be oxidized into nitrate (NO_3) by nitrifying bacteria, absorbed through roots and incorporated into the DNA or protein of a plant, eaten and pissed away by a cow or a human, broken back down to ammonia, oxidized into nitrate, built into a protein again. Sewage is loaded with biologically available nitrogen, which helped to fuel blooms of algae that plagued rivers like the Thames, the Hudson, and the Illinois.

In the early twentieth century, the activated sludge process, with its groomed populations of nitrifying bacteria, was a major improvement on existing ways of treating urban sewage. It was fast, and it didn't take up lots of space, as large contact filter beds did. Pumping air into the system used energy, but nobody was concerned about that at the time. Within a year of Ardern's presentation at the Society of Chemical Industry, fifteen cities around the world had begun testing the process and building sewage plants designed to put it into action.[46] By the early 1920s, activated sludge treatment plants were up and running in cities scattered across Britain and the US.[47]

The rapid rise of activated sludge was halted by a dispute over patents. Gilbert Fowler had quietly sold rights in the activated sludge concept, developed by his underlings Ardern and Lockett while all three were employed by the City of Manchester, to Walter Jones, an entrepreneur and ironworker. Jones had been among the first to work out ways to scale up the activated sludge process from laboratory bottles to industrial-sized tanks that could handle city sewage. His firm, Jones & Attwood Ltd., took out British patents in 1914, laying claim to both the idea and the hardware involved in activated sludge treatment. "The Sewage Work of the World is a big thing," wrote J.A. Coombs, Jones & Attwood's chief sewage engineer, "and the firm are by no means selfish in trying to corner it."[48] Jones & Attwood warned American engineers against infringing on the firm's activated sludge patents. The American Society of Civil Engineers organized to prevent the grant of a patent in the US, but their appeal failed.

Meanwhile, more US cities began to test and build activated sludge systems. As Chicago's population boomed, the limitations of its existing system of routing raw sewage into the Illinois Valley became obvious. Chicago pumped increasing amounts of water from Lake Michigan to dilute the waste in the Sanitary and Shipping Canal. Neighboring cities on the shores of Lake Michigan sued to protect their share of the lake's waters. The case moved through the legal system all the way to the US Supreme Court, which issued a ruling limiting the amount of water Chicago could take from the lake.[49] The need for sewage treatment then became urgent. The city began building and testing activated sludge plants in the early 1920s.

In 1924, Activated Sludge Inc. (ASI), the US representative of Jones & Attwood, filed suit against the city for patent infringement. Chicago forged ahead. The lawsuit languished in the legal system for years before any real action was taken. In 1927, Chicago put its North Side Treatment plant, the first full-scale activated sludge system in the city, into operation. The plant, designed to treat 175 million gallons of sewage per day, is still in use nine decades later.

ASI next went after the city of Milwaukee, Wisconsin, farther north on the shore of Lake Michigan. Both cities drew their drinking water from the lake and had a long history of dumping raw sewage into it, with serious consequences for public health. Milwaukee's Jones Island activated sludge plant started up in 1925, and the same year ASI filed suit for patent infringement.

The Milwaukee case went to trial in June 1928. Fowler and Ardern testified before Judge Ferdinand Geiger that the concept of activated sludge treatment had been developed in the City of Manchester's laboratory, under Fowler's supervision. Yet Judge Geiger bought into the fiction, supplied by ASI's attorney, that both the scientific experiments and the equipment needed to expand the concept to an industrial scale had leapt from Walter Jones's mind.[50] Jones could not comment: He had died the year before.

In 1933, Geiger found the city of Milwaukee liable for hundreds of thousands of dollars in damages to ASI. He also issued an injunction that would have forced the city to shut down its treatment plant, causing the release of raw sewage to Lake Michigan. On appeal, Geiger's decision was upheld except for the injunction against operation of the Jones Island plant. In the view of the appeals court, the threat to public health outweighed the infringement of ASI's patents. In the fall of 1934, after the Supreme Court had declined to hear an appeal of Milwaukee's case, the judge in the Chicago lawsuit issued his decision. He seemed skeptical that Walter Jones had been the sole inventor of the activated sludge process, but he found the patents to be valid nonetheless. In the end, after more than two decades of litigation, Chicago was forced to pay ASI $950,000. Milwaukee paid $880,000.

At the time of the appeals court decision, in 1937, *Time* magazine reported that several other US cities had paid lump-sum license fees to ASI to avoid the long, expensive process of a court case. The fees ranged from $85,000 for Cleveland and $75,000 for Houston to $23,000 for Peoria, Illinois. ASI had lawsuits pending against New York and more than one hundred other cities in the US.[51] Most had refused to pay, arguing that they'd designed their own systems based on original experiments.

The outcomes of the Chicago and Milwaukee cases affected plans for treatment plants throughout the country. Several existing plants, including one in San Marcos, Texas, which in 1916 had become the first operating activated sludge system in the US, closed down to avoid fines. Others paid royalty fees to ASI in order to keep their plants running. A large number of communities that had plans to build new activated sludge plants used an alternative—usually a trickling filter system like the one Dibdin had built at Barking—or just sat back and waited.

Los Angeles, which had been the target of an ASI lawsuit for patent infringement, gave up on the idea of an activated sludge treatment plant and did not go beyond installing filters to catch the larger solid chunks before raw sewage was discharged to the Pacific. By the 1940s, acres of sewage could be seen floating offshore. Clots of dung and debris settled on the beaches. In the summer of 1943, the State Board of Health quarantined Los Angeles' beaches because of extremely high levels of *E. coli* in the water.[52] When people kept coming, Dr. Elmer Belt of the Board of Health advocated arresting bathers and beachcombers for their own protection. The city's first activated sludge plant was not built until 1950, and within a few years it had failed to keep up with the ever-growing population and volume of sewage.

A young man came of age in those years. He grew up not far from Los Angeles Harbor, where the water was so drained of oxygen that only a single hardy species of marine worm could survive. His name was David Joseph, and he would use the weapons of science to wage war on "pollution as usual" in California. In time he would clash, long and hard, with Arcata's sewage rebels.

NOTES

1. Anonymous (1858). "The Thames in his glory." *Littell's Living Age* **58**: 375.
2. Hamlin, C. (1980). "Sewage: waste or resource?" *Environment* **22**(8): 16–42.
3. Halliday, S. (1999). "The Great Stink of London: Sir Joseph Bazalgette and the cleansing of the Victorian metropolis," p. ix.
4. Hamlin, C. (1987). "What becomes of pollution? Adversary science and the controversy on the self-purification of rivers in Britain, 1850–1900."
5. Hamlin, C. (Spring 1985). "Providence and putrefaction: Victorian sanitarians and the natural theology of health and disease." *Victorian Studies.*
6. Jacob, A. (1870–1871). "The treatment of town sewage." *Minutes of Proceedings, Instituion of Covil Engineers* **32**: 402–404.
7. Hamlin, C. (1987). "What becomes of pollution? Adversary science and the controversy on the self-purification of rivers in Britain, 1850–1900," Chapter IV.
8. Ibid., p. 173.
9. Darwin, C. "Autobiography of Charles Darwin, 1809–1882," as quoted in C. Hitchens, "The portable atheist," p. 94.
10. Hamlin, C. (1988). "William Dibdin and the idea of biological sewage treatment." *Technology and Culture* **29**(2): 189–218.
11. Hamlin, C. (1987). "What becomes of pollution? Adversary science and the controversy on the self-purification of rivers in Britain, 1850–1900," pp. 52–53.
12. Hamlin, C. (1988). "William Dibdin and the idea of biological sewage treatment." *Technology and Culture* **29**(2): 189–218.
13. Hamlin, C. (1980). "Sewage: waste or resource?" *Environment* **22**(8): 16–42.
14. Halliday, S. (1999). "The Great Stink of London: Sir Joseph Bazalgette and the cleansing of the Victorian metropolis," p. 109.
15. Hamlin, C. (1987). "What becomes of pollution? Adversary science and the controversy on the self-purification of rivers in Britain, 1850–1900," p. 36.
16. Hamlin, C. (1980). "Sewage: waste or resource?" *Environment* **22**(8): 16–42.
17. Ibid.

18. Hamlin, C. (1987). "What becomes of pollution? Adversary science and the controversy on the self-purification of rivers in Britain, 1850–1900," pp. 490–491.
19. Sorby, H.C. (1884). "Detection of sewage contamination by the use of the microscope, and on the purifying action of minute animals and plants." *Journal of the Royal Microscopical Society* **4**: 988–992.
20. Hamlin, C. (1987). "What becomes of pollution? Adversary science and the controversy on the self-purification of rivers in Britain, 1850–1900," p. 505.
21. Hamlin, C. (1988). "William Dibdin and the idea of biological sewage treatment." *Technology and Culture* **29**(2): 189–218, p. 212.
22. Schneider, D. (2011). "Hybrid Nature." p. 8. *Cambridge, Massachusetts and London, England: The MIT Press.*
23. Hamlin, C. (1987). "What becomes of pollution? Adversary science and the controversy on the self-purification of rivers in Britain, 1850–1900," p. 518.
24. Schneider, D. (2011). "Hybrid Nature," p. 30.
25. Thudichum, G. (1897). "The ultimate purification of sewage." *Engineering*: 192.
26. Anonymous (July 16, 1897). "Editorial." *Baltimore News.*
27. Schneider, D. (2011). "Hybrid nature," p. 10.
28. Jordan, E.O. (1890). "On certain species of bacteria observed in sewage." In "A report of the biological work of the Lawrence Experiment Station, in Experimental Investigations by the State Board of Health of Massachusetts, upon the purification of sewage by filtration and by chemical precipitation and upon the intermittent filtration of water. Made at Lawrence, Mass, 1888–1890," part V.
29. Hill, L. (2000). "The Chicago River: a natural and unnatural history," p. 121.
30. Schneider, D. (2011). "Hybrid nature," p. xxiii.
31. Sedgewick, W., Allen Hazen (April 21, 1892). "Typhoid fever in Chicago." *Engineering News and American Railway Journal* (accessed at http://www.encyclopedia.chicagohistory.org/pages/10722.html).
32. McCarthy, M.P. (1993). "Should we drink the water? Typhoid fever worries at the Columbian Exposition." *Illinois Historical Journal* **86**(1): 2–14.
33. Sedgewick, W., Allen Hazen (April 21, 1892). "Typhoid fever in Chicago." *Engineering News and American Railway Journal* (accessed at http://www.encyclopedia.chicagohistory.org/pages/10722.html).
34. McCarthy, M P. (1993). "Should we drink the water? Typhoid fever worries at the Columbian Exposition." *Illinois Historical Journal* **86**(1): 2–14.
35. Hill, L. (2000). "The Chicago River: a natural and unnatural history," p. 123.
36. Ibid., p. 134.
37. Ibid., p. 132.
38. Jordan, E.O. (1901). "The relative abundance of *Bacillus coli communis* in river water as an index of the self-purification of streams." *Journal of Hygiene* **1**(3): 295–320.
39. Leighton, M.O. (1907). "Pollution of Illinois and Mississippi Rivers by Chicago sewage: a digest of the testimony taken in the case of the *State of Missouri v. the State of Illinois and the Sanitary District of Chicago." Bureau of Reclamation, Government Printing Office.* Accessed at http://pubs.usgs.gov/wsp/0194/report.pdf, p. 207–240.
40. Williams, D. (March 9, 1995). "Shifting the burden in pollution cases." *St. Louis Post-Dispatch*, p. 7B.
41. Tarr, J., Francis Clay McMichael (October 1977). "Historic turning points in municipal water supply and wastewater disposal, 1850–1932." *Civil Engineering*, pp. 82–86.

42. Hill, L. (2000). "The Chicago River: a natural and unnatural history," pp. 134–135.
43. Schneider, D. (2011). "Hybrid nature," p. 31.
44. Ardern, E., W.T. Lockett (1914). "Experiments on the oxidation of sewage without filters." *Journal of the Society of Chemical Industry* **33**(10): 523–539.
45. Ackert, L. J. (2006). "The role of microbes in agriculture: Sergei Vinogradskii's discovery and investigation of chemosynthesis, 1880–1910." *Journal of the History of Biology* **39**(2): 373–406.
46. Schneider, D. (2011). "Hybrid nature," p. 34.
47. Alleman, J. E., T.B.S. Prakasam (1983). "Reflections on seven decades of activated sludge history." *Journal of the Water Pollution Control Federation* **55**(5): 436–443.
48. Schneider, D. (2011). "Hybrid nature," p. 53.
49. Janicke, P.M. "Wastewater treatment patent controversies: the unsettled meaning of 'inventor.'" *Spirit Over the Waters: Two Legal History Libraries About Water*, **Library 2** (http://www.watercases.org/LIB_2/Story%20of%20Library%202%20--%20P.%20Janicke.pdf).
50. Ibid.
51. Anonymous (July 5, 1937). "Activated Sludge, Inc." *Time*, pp. 48–49.
52. Anonymous (July 15, 1943). "Basis of beach quarantine told by state experts." *Los Angeles Times*.

4

Emperor Joseph's Roots

On a May morning in 1957, ten thousand fish floated on the eastern edge of San Francisco Bay, their pale, upturned bellies bobbing on the surface of the dark water. The crowd of carcasses described an arc that stretched along the shore from Richmond's harbor south to Point Isabel. Many striped bass, a prized game fish, were among the dead.

Seth Gordon, director of California Department of Fish and Game (DFG), fielded complaints from anglers outraged by the fish kill. The Public Health Committee of the State Assembly passed a resolution admonishing DFG for its failure to enforce pollution control laws. Gordon told the committee members off. "We want to stop pollution," he said, "but the law as it stands puts our Department in the position of a boxer going into the ring with one hand tied behind his back."[1] The ability to set and enforce pollution standards rested with California's nine regional water pollution control boards. To effect any change, Gordon's department had to prove to the boards' satisfaction that pollution allowed by existing standards was harmful to fish, a challenge that had so far proved impossible. Responding to questions about the East Bay fish kill, he said, "We still don't know what caused the die-off, or where it came from."

David Joseph was then starting out as a DFG biologist, armed with a doctorate in marine biology from the University of California at Los Angeles. Born in Connecticut, on a cooperative farm where his parents raised dairy cows and shade-grown tobacco with other immigrant Russian Jews, he'd grown up in Inglewood, in southern California, when the place was still a bucolic town and he could ride his horse to the beach. He'd met his wife, Marion, when they were both students at UCLA. "He was an outdoor guy," she remembers. "He wasn't a fisherman, he just loved the sea, loved the land. His work was always going to have something to do with protecting the environment."

In 1957, Joseph was an oddity—at the time, few, if any, DFG staffers held graduate degrees. In the aftermath of the East Bay fish kill, he pushed his colleagues to nail down the cause of the disaster, using careful scientific detective work.

At Joseph's urging, DFG staff tallied the numbers of dead fish, brought some of the victims back for study, and collected samples of the water for analysis. The mass of fish carcasses had centered around the Stauffer chemical plant in

Richmond. Analyses of both bay water and the dead fish found significant levels of toxins released in Stauffer's effluent. DFG biologists checked the chemical content of discharges from other nearby plants, and concluded the deadly chemical could have come only from Stauffer. The gills of the dead fish had been eroded away, causing them to suffocate. Laboratory tests found that fish exposed to the effluent from the Stauffer plant died. Their gills were destroyed, just as the gills of the fish in the East Bay kill had been; under the microscope, the same pattern of damage was clear.

Fed up with the determined inaction of the San Francisco Water Pollution Control Board, Joseph suggested a new tactic: The state could sue Stauffer for damages. It took time to gather evidence and to convince his superiors the lawsuit was a valid idea. In March 1958, the state attorney general (Edmund G. "Pat" Brown, soon to win his first term as California governor) filed a civil lawsuit against Stauffer, seeking $13,000 in damages, the price tag DFG put on the more than two thousand striped bass that had died.

Newspaper reports described the suit as "unprecedented." Stauffer's attorney branded the suit "ridiculous" and announced that Stauffer's effluent was "incapable of destroying fish." By the time the trial was under way, Stauffer had hired a new lawyer, who acknowledged that the company discharged twenty-five tons of acid to the bay each month. He argued that the fish kill could not be Stauffer's fault, because no self-respecting fish would swim into water as befouled as that near the Stauffer outfall. Joseph testified as an expert witness and made sure that the jury saw photomicrographs of the eroded gills in affected fish. In December 1959, a jury awarded the state more than $15,000 in damages from Stauffer—funds that would go to the DFG. Today that amount seems puny, but at the time California industrialists found it outrageous, and feared that it set an ominous new standard for environmental protection.

The Stauffer case marked the start of a long career in which Joseph would use a combination of scientific rigor, personal charm, and sheer stubbornness to shake up inert regulatory agencies, to subvert the "politics as usual" that allowed ongoing water pollution. One colleague described Joseph's strategy as "poking the bear." Others saw him as a David battling the Goliath of industrial pollution.

Arcata's activists thought of Joseph as an arrogant scion of the state bureaucracy. Irked by his refusal to take them or their arguments seriously, they called him "Emperor Joseph." Yet many of those who knew him well viewed Joseph as a groundbreaking environmental hero. Perhaps he was both: a clean-water revolutionary who came to resist innovative, but in his view unproven, solutions.

Joseph began working for the state in an era when industry and city governments assumed that dumping wastes in rivers, bays, and the ocean was their right, and the regulatory boards charged with protecting California's waters agreed. In 1949, the California Legislature had passed the Dickey Water Act, setting up nine regional water pollution control boards. Board members represented agriculture, industry, cities, and counties. They shared an interest in maintaining their right to pollute.[2]

"The use of the State's waters for waste disposal in order to provide a better economic position for a municipality or an industry has become a guiding philosophy in the application of the Dickey Act," Gordon testified at a 1957 legislative hearing.[3] Under the law, he noted, pollution did not legally exist until it reached a level determined to be "unreasonable." While anglers and biologists saw unreason in tainted waters all around the state, the regional water pollution control boards saw an orderly use of resources, with certain stretches of water designated as official industrial or municipal cesspools. Gordon suggested some changes in state law that were, at the time, radical. No public waters should be completely set aside for waste disposal. Penalties for pollution should be made so high that no plant on shore could afford to dump raw effluent into California's bays.

At the start of his DFG career, Joseph's office was at San Pedro, close enough to the intensely polluted Los Angeles–Long Beach harbor that he could smell the rotten-egg stink that rose from the water. The harbor received effluent from fish canneries, oil refineries, chemical plants, vegetable oil plants, metalworking shops, shipyards, and city sewage outfalls.[4] In dry summers, Dominguez Channel, a concrete-lined canal that ran into the harbor from the north, carried no liquid aside from the waste discharges of industries along its banks.[5]

The dense kelp beds that had once flourished in open waters outside the harbor's mouth thinned out, and then vanished—as they did outside of polluted San Diego Bay, and for miles near a major urban sewage outfall at White's Point on the southern California coast, where discharge rates had risen from 20 million gallons a day in 1940 to more than 200 million gallons a day in 1957.[6] The abalone and other creatures that relied on kelp for food or shelter also dwindled. On beaches near major sewage outfalls, kids played with the rubber rings from decayed condoms, which washed up in the thousands.

Inside Los Angeles harbor, biologists were just beginning to explore the impacts of the intense pollution. Using a bucket equipped with retractable flaps that could scoop up samples of the bay floor, Donald Reish, a graduate student in marine biology at the University of Southern California, found that the creatures living on the harbor floor—the benthos—were profoundly affected by contaminants. Reish focused his study on polychaetes, segmented marine worms that were among the most common creatures found in the harbor. He identified six kinds of benthic communities, defined by the level of dissolved oxygen in the water and the diversity of polychaete species. (Pollution with organic wastes and petroleum tends to deplete dissolved oxygen [DO], suffocating marine life.) Mapping the harbor from a worm's point of view, Reish defined areas that ranged from a healthy bottom, with a diverse array of polychaetes and a normal DO level of about 7 parts per million (ppm), to a polluted bottom in waters where the DO was 3.5 ppm, about half the normal level, and the community was dominated by a single hardy species of worm, *Capitella capitata*. In areas he dubbed "very polluted," the bottom was covered with a layer of toxic sludge in which nothing could survive. Brought to the surface, this black muck stank of sulfur and petroleum.[7]

Fish captured in or near the harbor showed signs of serious health problems. Halibut were underweight and listless; white seabass and spotfin croaker had exophthalmia, an abnormal protrusion of the eyes, leading to blindness and death; a variety of fish had mysterious lesions and tumors. In an experiment, DFG biologists put wild killifish into effluent collected from Dominguez Channel and diluted with seawater. The fish developed lesions within twelve days, and all of them died, even those that were moved into clean water.[8]

Engineers at the Los Angeles Harbor Department preferred foul water. In clean conditions, marine worms burrowed into the thousands of wooden pilings in the harbor, and hundreds of the weakened pilings had to be replaced each year. Carrol Wakeman, one of the harbor engineers, sang the praises of pollution, telling a legislative committee:

> The more pollution we have in the harbor, the fewer pilings need to be replaced. It also saves money for boat owners; where there is pollution they don't have to paint their boats as often, because barnacles and other fouling mechanisms are nonexistent. This is very important to us . . . We don't want objectionable floating matter but we do want the type of pollution which limits oxygen content in harbor waters.

Wakeman said he'd be happy if the entire inner harbor had oxygen levels of about 2 ppm, because the waters would be conveniently dead. When a stunned assemblyman commented that Wakeman was the first person he'd ever encountered who liked pollution, Wakeman replied, "It's a matter of dollars and cents—economics."[9]

The waters of San Diego Bay were tinted a brownish red, the result of intense blooms of algae fueled by mass discharges of raw or minimally treated sewage. The tiny, floating marine plants would take up the nutrients in the sewage and reproduce like mad; at night, when photosynthesis was impossible, the algae would respire, breathing in oxygen just as mammals do. As masses of phytoplankton died off, they sank to the bottom, where bacteria used up more oxygen in the process of decomposing them. As a result, the bay's waters were drained of oxygen, suffocating fish and marine invertebrates. The polluted southern arm of San Francisco Bay likewise went anoxic every summer, the result of discharges of fish cannery wastes and poorly treated city sewage.[10]

The Los Angeles Regional Water Pollution Control board designated disposal of industrial wastes and sewage as a "beneficial use" of the Los Angeles harbor, along with the provision of docking facilities for ships. Habitat for marine life didn't make the list. Joseph came on the scene just as citizen's groups and his own new employer, DFG, were taking up the fight against rampant pollution. He would soon prove his ability to think—and to fight—outside the box. Jack Fraser, director of water projects for DFG, created a new position for Joseph, who in the late 1950s became California's first full-time water quality specialist. Joseph moved, with his wife and young kids, to Sacramento, and began to tackle water pollution issues statewide.

He would spend the next few years in pitched battle with the regional water pollution control boards. Board members were appointed by the governor, and as prescribed by the Dickey Act, were linked to industry, agribusiness, and city governments, all of which had a need to dump effluents containing everything from human sewage to caustic solvents and toxic waste from oil refining and the manufacture of DDT. The regional boards focused on making life easy for polluters, and stubbornly ignored protests from DFG staff, conservationists, and fishermen.

While DFG Director Gordon and his successor, Walt Shannon, continued to advocate for change in the Dickey Act, Joseph searched for ways to move ahead under existing law. By 1961 Joseph had helped to push through another legal case, in which Shell Oil was charged with criminal pollution of Dominguez Channel. While the Shell case was pending in municipal court in Compton, Joseph and his colleagues moved to gather evidence on the dire biological state of Dominguez Channel and the inner harbor. Frequent kills occurred when schools of fish from the cleaner outer harbor blundered into the inner harbor and suffocated. Joseph hoped to do an end-run around the inert Los Angeles Regional Board and use the courts to enjoin polluters from discharging untreated waste. The strategy relied on language in the DFG Code section 5650, which seemed to give the DFG the power to act independently of the regional boards. DFG enlisted the help of Ralph Scott, the assistant attorney general who'd prosecuted the case against Stauffer. Scott thought the agency had solid grounds to seek injunctions against the industrial plants that lined Dominguez Channel.

In July 1961, Joseph went into high gear, gathering evidence on fish kills, assigning his staff to conduct bioassays of industrial effluent and necropsies on killed fish in the harbor. He studied the work of Don Reish, who'd identified a series of dead zones on the floor of the Los Angeles harbor in 1954. He was hard at work building a case against the Dominguez Channel polluters when, in November, the judge in Compton dealt DFG a stunning blow, dismissing the agency's case against Shell before it could be decided by a jury. The judge said that Dominguez Channel did not constitute "waters of the state of California" because in the dry months it carried nothing but the effluent flowing from dischargers. He refused to allow expert testimony on the toxicity of inner harbor waters to fish, or to recognize DFG's right to act independently of the regional board. In the aftermath of this defeat, DFG Director Walt Shannon wrote to Ralph Scott asking him to move ahead with case-by-case injunctions against industrial polluters.

Scott replied that any hope for the DFG cases in Dominguez Channel rested on the outcome of an appeal then pending on a case concerning pollution in Calaveras County, on the Mokelumne River. Months later, the appeals court ruled that injunctive relief in pollution cases was the exclusive province of the regional water pollution boards. DFG's legal maneuvers had failed.

The Los Angeles Regional Board stood by its decision to do nothing. The board had declined to set discharge requirements for individual polluters; instead it set a water quality objective that DO in inner harbor waters should not fall below 0.5 ppm, a peculiar goal since most forms of marine life can't survive in water that oxygen-poor. In 1956, DFG Director Seth Gordon sent a complaint to the regional board, noting that pollution levels in the inner harbor had risen, with obvious effects on fish. When the waste flow from the Richfield Oil plant was diverted from the Los Angeles County sewer system back to Dominguez Slough, hundreds of gulls mobbed the harbor, gobbling up dying anchovies. A valuable species of bait fish that was once abundant in the harbor, anchovies go to the surface and show obvious signs of distress when DO levels drop below 3 ppm.[11]

In the fall of 1963, the same kind of disaster played out again on a larger scale: Millions of anchovies died in a kill that attracted an estimated 20,000 gulls. The dying anchovies made such easy prey that the gulls ate until they were too heavy to fly. DFG Captain Walter Putnam pointed out that this kind of thing had been happening in the Los Angeles harbor for years.[12] More than half of the bait fish taken each year in California were anchovies captured in the outer harbor. He worried that the outer harbor would one day be as polluted as the inner harbor, a catastrophe that he speculated could put an end to sport fishing in southern California, because it would destroy the principal source of bait. In places the water turned a pearlescent white, an effect created when sulfates discharged from oil refineries fed blooms of anaerobic bacteria. White masses of colloidal sulfur were left behind. Putnam regularly tracked oil spills in the harbor, sometimes several in a single day, and watched as a cannery discharge bubbled a brown slop of fish bones and guts into the water.

The fight for the Los Angeles harbor dragged on for years. In late 1958, during the closing weeks of his last term as governor, Pat Brown appointed Ellen Stern Harris, a dedicated conservationist, as the public's representative on the Los Angeles Regional Board. Harris questioned every move her fellow board members made, drawing public attention and slowing down the board's normal rubber-stamping of pollution permits. Long-time board members snapped at her, asking her with varying degrees of rudeness to shut up and get out of the way. Thomas Gaines, a public relations officer at one of the Dominguez Channel oil refineries, found her particularly annoying. In 1967, when officials with the Federal Water Pollution Control Administration admonished the Los Angeles Regional Board for failing to monitor and regulate DO levels in the harbor and to acknowledge fish life as a beneficial use, Harris urged her fellow board members to change their standards. By then, Harris's resistance had drawn activists to the regional board's meetings, and other conservationists were on hand to back her up. Gaines described the push for higher water quality standards as "an exercise in futility."[13] The board voted to change nothing. The majority shared Gaines' conviction that California water quality was none of the Feds' business.

Conservationists, along with DFG, kept pushing. Demonstrators took a flotilla of small boats into the harbor and dropped black wreaths into the water in protest. Angry editorials ran in the *Los Angeles Times* and other local papers. By late 1968 DFG had a new case in court, involving pollution of Dominguez Channel by Union Oil Company. Two appellate courts had found against the company's contention that DFG could not enforce its code because it had been superseded by the power of the regional boards; a third appeal by Union Oil was pending.

The clash between industry and conservationists played out at a hearing held by the Los Angeles Regional Board in April 1969. A Sierra Club representative read historical accounts that proved the now-sterile harbor had once been rich in marine life, and urged the board to adopt fish and wildlife habitat as a beneficial use of the harbor. John Easthagen, a chemical engineer and spokesman for the Western Oil and Gas Association, replied that the idea of bringing life back to the harbor was "an irrational, emotional and unscientific approach."[14] The harbor and Dominguez Channel, he argued, were manmade structures, and not intended for marine life.

Easthagen was right, in a way, about the history of the Los Angeles harbor. Until the early twentieth century, the shoreline of southern Los Angeles County was covered in salt marsh and mudflat. Dominguez Slough was a mucky, meandering channel that flowed in from the north to meet the mouth of the Los Angeles River. In the early 1900s the marshes were drained, the Los Angeles River was diverted away, and wharves and a breakwater were built to allow deep draft ships into the harbor. By the 1950s, Dominguez Channel was a concrete canal running through a paved landscape, fed by manmade storm drains that carried runoff from 110 square miles of southern Los Angeles County, including Dave Joseph's childhood home of Inglewood.

Bringing these urbanized, industrial waters back to life, Easthagen claimed, would be "very costly, of doubtful value . . . unreasonable, arbitrary and capricious." Ellen Stern Harris told him that the harbor area belonged to the public, and the people now wanted to reclaim it. "I don't think you can go back to the time before the Indian," Easthagen answered.

The majority of the Los Angeles Regional Board agreed with Easthagen, and held to their traditional stance of doing nothing to regulate industrial discharges. But change was coming, whether the regional board liked it or not. At a September hearing, conservation-minded members of the State Water Resources Control Board raked the Los Angeles Regional Board over the coals for its inaction. Kerry Mulligan, the state board's new chairman, told them the day might come when all discharges into the inner harbor and Dominguez Channel would be banned. At the end of the meeting, Lester Louden, the long-time chair of the regional board, went up to Jerome Gilbert, the state board's executive officer. He suggested that Gilbert change his name to "Ellen Stern Gilbert," a mocking reference to activist Ellen Stern Harris. A few days later, Louden succumbed to pressure from the state board and resigned his chairmanship. The regional board grudgingly agreed to list marine life as a beneficial use of the harbor.

In October 1969, the state board ordered an end to all waste dumping in Dominguez Channel and the inner harbor, a change that was to be completed by 1973. Industrial effluents would have to be treated and discharged to the ocean rather than the inner harbor. It was a seismic shift in policy, brought on by years of struggle. And it worked: by November 1970, inner harbor waters had gained back some oxygen—2 ppm—and fish appeared in the murk. By February 1971, schools of anchovy and bonita had returned, and so had the barnacles and wood-boring polychaete worms that clung to the wooden pilings in the Los Angeles harbor. A professional diver who'd worked in the harbor for years, feeling his way in water so turbid he couldn't spot his hand in front of his face, reported that he could see where he was going in the depths of the harbor for the first time ever.

The decade between the dismissed case against Shell Oil and the rebirth of the Los Angeles harbor had been a busy one for Dave Joseph. He'd ended up in a position of power that no one, not even his closest allies, had thought he could achieve. Joseph's unexpected rise was tied to a long fight over pollution from pulp mills, then a new industry in California.

In the spring of 1960, the North Coast Regional Water Pollution Control Board was poised to approve the release of effluent from a pulp mill that the Simpson Timber Company planned to build at Fairhaven, on the Samoa Peninsula, west of Humboldt Bay—the first pulp mill proposed on the California coast. Simpson planned to discharge its effluent, containing a heavy load of organic waste and caustic chemicals, directly onto the beach on the Pacific Ocean side of the peninsula. This strategy had already proved disastrous at a Georgia-Pacific pulp mill on the Oregon coast, near Newport, where effluent released onto the beach had created a foul-smelling wall of foam that triggered a wave of outraged protest from local people.

In the 1960s, as many lumber mills shut down, the wood pulp industry held the promise of new jobs in hungry logging towns like Arcata and Eureka. (The Humboldt Chamber of Commerce recorded 262 working sawmills in the county in 1951 but only fifty-five by 1962.) But the economic boost came at a price. To create a smooth pulp that yields strong paper, wood chips are cooked in an alkaline soup, under pressure. The process produces large volumes of waste that can be deadly to aquatic life. Loaded with wood fibers and lignin, a component of plant cell walls that gives wood its structural strength, the effluent has a high biochemical oxygen demand—that is, in the process of decomposing, it sucks dissolved oxygen out of the water. It also contains mercaptans (carbon-based sulfur compounds), acids, and sulfides, all of which can irritate the gill surfaces of fish and impair the growth of oysters. High doses of methyl mercaptans can kill fish by paralyzing the nerves of their gill muscles.[15] (Mercaptans cause the infamous stench associated with pulp mills. During the boom years of pulp production in Humboldt County, prevailing winds carried the stink inland from the Samoa Peninsula to Eureka, causing the whole city to hold its collective nose.)

The North Coast Regional Board was as reluctant to demand expensive environmental protections from industry as its Los Angeles counterpart. William Shackleton, the regional board's executive officer, sent Simpson a list of wastewater "requirements" that were so general as to be meaningless. Joseph protested. DFG and other agencies were supposed to comment on the board's standards, but they'd been given no time to study the issue and no information beyond a map showing the location of the mill site and the fact that Simpson expected to release 30 million gallons of liquid waste every day. Shackleton blandly ignored pleas for a delay from Joseph, and from the Departments of Public Health and Water Resources.

In July, DFG hosted a meeting in Eureka to discuss pollution from Simpson's proposed mill. They invited representatives from all the involved state agencies, local fishermen's and conservation groups, and Simpson. Shackleton responded to his invitation with a letter announcing that no representative of the regional board would attend as the agency had nothing to contribute.

During the meeting, several people objected to the notion of releasing effluent onto the beach. George Allen, a fisheries professor at Humboldt State, worried that noxious effluent would affect surf fishing, and that wastewater discharged to the ocean might make its way into Humboldt Bay on the incoming tide. Joseph found other allies at the meeting. One was Edward Eldridge, a scientist with the US Department of Public Health based in Oregon, which had been grappling with pulp mill discharges for decades. Eldridge agreed it was obvious an outfall pipe should be built to carry the discharge well away from the shore. He mentioned the foam-drowned beach at Newport.

Roland Sultze, the Simpson rep, said one of the major draws of the Fairhaven site on the Samoa Peninsula was the ability to discharge directly to the ocean. Discharging to rivers was just too much trouble—decades of experience had shown how devastating the effluent could be to fish and to the look and smell of natural waters. Sultze admitted to Joseph that the regional board's vague regulations would make it difficult to plan and engineer the mill. His closing comment was that if pulp mills could not dispose of wastes in the Pacific Ocean, they wouldn't be built in California at all.

In November 1960, Joseph traveled north with a group of regulators and industrialists involved in plans for a pulp mill on the Sacramento River. They visited a series of mills in Oregon and Washington, all of which, Joseph observed, had serious pollution problems in their early years. "Improvements came about," he wrote, "only after a considerable amount of public indignation." In Oregon, as in California and everywhere else, industry would only go as far as it was pushed in terms of investing cash in waste disposal systems. On the Willamette River, where it ran through Portland, areas with low DO levels created by sewage discharges and mill wastes at times formed barriers to migrating salmon; to continue upstream meant suffocation.

At the Crown-Zellerbach mill on the north bank of the Columbia River, he learned the details of the company's struggle with mats of pollution-fed algae building up in the river. These clumps of "slime" fouled fishing gear and brought complaints from local people. Rather than treating its effluent, Crown-Zellerbach built holding ponds so that its discharge could be released gradually, avoiding intense algal blooms. The Crown-Zellerbach mill smelled awful, an inevitable part of industrial wood pulping. Joseph mused that the healthiest attitude toward a new mill was a balance of resignation and vigilance. Economic forces meant the mills would come; the resulting changes would be sad, even if regulators did the best they could.

Dale Snow of the Oregon Fisheries Commission led Joseph on a tour of the Georgia-Pacific mill that had become infamous for discharging onto the beach at Newport. The history of the mill's effluent had a familiar ring. Before the plant was built, Georgia-Pacific planned to dump its waste into Yaquina Bay, the home of an extensive oyster farming industry. Oregon regulators refused to allow this, and a battle went on for more than a year while the company tried to avoid the added expense of piping its effluent past the bay and out to sea. At last Georgia-Pacific built an eight-mile-long pipe that carried its effluent onto the beach in front of the town of Newport. There it gurgled onto the sand, creating a bank of foam that stank to high heaven. After many months of complaint from the citizens of Newport, Georgia-Pacific had extended its discharge pipe 250 feet into the ocean, and the worst of the foam and odor problems subsided. Snow was tracking impacts on nearshore fish and invertebrates, and had so far found none.

Joseph returned from Oregon convinced that a mill can do anything that's required to prevent water pollution, if the owner is made to invest the money. A scientist at the Crown-Zellerbach mill had pointed out that pulp mill wastes could be treated in the same way as domestic sewage, and that this was already being done at four US mills. "They will do whatever they are forced to do," Joseph wrote. "DFG and the Water Pollution Control Board will force the company to do whatever is necessary."

At Oregon State College's pollution lab, which was studying the effects of pulp waste on salmon, he'd met Bob Lewis, a young scientist who soon after came to work for DFG as a pollution analyst. "Dave was one of the first to grapple with the hierarchy to try to get more protection for fish and wildlife," remembers Lewis. "He had a great sense of humor, bright as a new dollar, very compassionate about other people. We called him Little Giant, but not to his face. On the north coast he was known as Dr. Clean; the dischargers came up with that name."

Based on information from existing mills and advice from Erman Pearson, a University of California professor recognized as an expert on pulp mill waste disposal, Joseph put together recommended regulations for the Simpson mill. Half a century later, his work still looks impressive. His guidelines addressed subtleties that were then often ignored. Impacts on fish were (and often still are) measured by exposing fish in a laboratory to varying dilutions of effluent for a few days and finding the amount that caused half the population to die, a metric known as $TL_{m,}$ for Threshold Limit, median. By the 1960s, biologists had begun to point out the obvious flaws in using this live-or-die standard. For wild fish populations to thrive, they need conditions in which all life stages, from eggs to adults, can stay healthy. At the time, little was known about the toxic effects of the many ingredients of pulp waste. Joseph tried to err on the side of safety by stating that the concentration of effluent in the receiving waters must not exceed a tenth of the TL_m value as measured on a free-swimming stage of a resident salmonid.

He also addressed problems with oxygen depletion, required holding ponds to control excess effluent, and demanded a monitoring program to track impacts on marine life—a request considered radical at the time. All this, including a mathematical formula for determining effluent concentrations in the ocean, fit neatly on a single page. The recommendations were submitted to the regional board in February 1961. Then came the painful process of waiting and watching while the regional board found ways to bury them.

At the board's meeting in March, Humboldt Bay fishermen and oyster farmers showed up to read letters of support for Joseph's regulations. "It appeared as though the Board felt our case had merit," Joseph wrote. Shackleton, the executive officer, who had a long-established habit of brushing off DFG's concerns, suggested that the board wait until Simpson had a chance to study the proposal, so no action was taken.

Over the next few weeks, Joseph and his boss, Jack Fraser, tried to convince the regional board's engineer, Wendell Candland, to read their recommendations. Candland ignored them. ("Wendell was annoyed when DFG hired its own pollution analysts," remembers Lewis. "He spent most of his time visiting wineries, supposedly to investigate them for pollution. He was a double jerk.") On May 18, Shackleton sent a letter informing them that nothing would change, because Simpson, unsure if and when it would break ground on the mill, did not wish to discuss the matter.

Joseph asked that the board cancel its vague regulations for Simpson's proposed mill and wait until Simpson could provide specific plans. He reminded the board that the original regulations were set hurriedly because of Simpson's supposed need for speed. "This suggestion," Joseph remarked, "aroused something less than wild enthusiasm in Shackleton."

Later in June the *San Francisco Chronicle* reported that Georgia-Pacific planned to build a pulp mill on the Samoa Peninsula, adjacent to Simpson's land. Worried by the vision of multiple unregulated pulp mills belching waste onto the beach, Joseph began to think of taking the issue up the chain of bureaucratic command to the State Water Pollution Control Board.

At the regional board's September meeting, the two women members (known to history only as Mrs. Gordon Hadley and Miss Childs) staged a small rebellion. They refused to second a motion to continue the original regulations for Simpson. Mrs. Hadley pointed out that the board had never had an analysis of DFG's suggested requirements from their engineer. Candland replied that he had not taken the time since he knew that the board had no intention of altering its original regulations. He suggested that an analysis of DFG's recommendations by an impartial engineer would take a year and cost at least $1,500. The board members ordered him to prepare an analysis for their next meeting, three months off.

To Joseph, Candland's declaration that he'd done nothing about the Simpson issue over the past eight months was "an unprecedented admission of incompetence." He was sure the only intelligent critique Candland could produce would have to come via Simpson's own engineers, and he expected the regional board to do nothing over the following three months but find ways to avoid the substance of his objections to unregulated pulp mill discharge.

"Back in the 50s and early 60s, this sort of thing was common," says Lewis. "The argument for letting Simpson dump waste on the beach was that the mill might smell bad but it was the smell of prosperity."

The regional board duly rejected DFG's recommendations in December 1961. In January 1962, Walt Shannon, the director of DFG, lodged a formal complaint with the state board. "We do not believe that the sanctioning of a 30 million gallon per day discharge of waste directly onto the beach in the Eureka area will promote the orderly development of the pulp industry," he told board members. He pointed out that the Kimberley-Clark mill on the Sacramento River had accepted strict discharge regulations, and that Georgia-Pacific had announced they would voluntarily install an extended pipeline to carry discharges from their Samoa mill out into the ocean.

In March 1962 eleven state board members traveled to a hearing in Eureka where Joseph and his allies had a chance to speak their minds. Joseph explained why the regional board's approach was too vague, and how destructive a discharge onto the beach would be. The Humboldt Bay Fisheries Association, which had been formed in 1961 as a response to the issue of Simpson's proposed pulp discharge, sent a representative to protest the regional board's refusal to set meaningful pollution limits. The cards had clearly been stacked in Simpson's favor, he argued, long before anyone who understood the environmental impacts had weighed in.

After that meeting, the paper trail on the Simpson mill controversy vanishes—for two years, no newspaper articles, no memos from Joseph to his superiors. There's no record that the state board ever made a decision on DFG's appeal. Perhaps the bureaucracy let the issue go because Simpson had no clear plans as to when it would start construction. Then, in June 1964, the *Eureka Humboldt Standard* ran a front-page story: Crown-Zellerbach Corporation was partnering with Simpson to build a $45 million pulp mill at Fairhaven. Combined with Georgia-Pacific's mill, already under construction, this would mean thousands of new jobs for Humboldt County. Governor Brown made a statement congratulating the people of Humboldt.

By the time the pollution controversy reappeared in the local newspapers, in the late autumn of 1966, the players had changed in surprising ways. Dave Joseph was no longer the DFG biologist petitioning to protect marine creatures: He'd been appointed executive director of the regional board, ending Shackleton's long and industry-friendly reign.

"The whole game started to change," recalls Lewis. "Dave was the first non-engineer to become a regional board executive officer—a real shock at the time."

"It was astonishing that they appointed a biologist," agrees Don Reish.

Joseph could now call the shots. His longtime ally in the fight over Simpson's pulp mill, Frank Douglas of the Coast Oyster Company, held a seat on the board. The notion of discharging onto the beach was quickly discarded. Under Joseph's leadership, the regional board required a discharge pipe that would extend two thousand feet into the ocean, and a monitoring program to track effects on marine life.

The long, unwieldy pipe was damaged when Simpson workers tried to install it. Attempts to repair the line failed, so the company had to wait for delivery of new pipe and favorable weather and tides that would allow its placement. The mill was ready to start up, at least on a limited basis, to test the equipment. The mill manager, P.M. Schnabel, announced the start of the mill's operations to the press—but he neglected to ask the regional board's permission to discharge through a foreshortened pipe that would release effluent near shore.

Joseph warned the company that its plan to release effluent through the broken outfall pipe was a violation of state regulations. A Crown-Zellerbach executive named Lowell Clucas flew to Humboldt to meet with Joseph and the Humboldt County district attorney. He made no comment on whether the mill would follow the regulations.

On Thanksgiving Day, 1966, the Simpson pulp mill started up, and effluent flowed into the Pacific near the shore. Joseph called an emergency meeting of the regional board on December 1. Ray Welsh, a board member and commercial fisherman, confronted mill manager Schnabel. "For four lousy days, you guys had to start up and couldn't wait," he growled. Still, the board agreed to let Simpson keep running with its broken discharge pipe until January 15, or until the new pipe was installed, whichever came first. Even Jack Fraser, Joseph's former boss at DFG, agreed with the plan, though he emphasized the need for Simpson to get its new discharge pipe in as soon as possible.

On a night two weeks later, a batch of black liquor—the chemical broth in which wood chips were cooked to yield pulp—escaped into the effluent pipe instead of being drawn back into the mill to reclaim its contents. (Pulp mill engineers had created ways to reuse the sulfur and salt in black liquor for economic reasons, long before environmental laws came into play. Black liquor is normally boiled down, leaving many of the useful chemicals behind and reducing the amount of wastewater.) Dawn revealed a bank of foam on the beach, five feet high, stretching for more than half a mile north and south of the discharge point. Douglas, the oyster farmer and regional board member, waded into the bank of waste and took in the strange view. The temporary permit the board had issued Simpson at its emergency meeting had specified that no foam was allowed. The power to make rules, even when you've won it, is not always enough.

In his Santa Rosa office, two hundred miles from the mess on Samoa's beach, Joseph told the press that the incident showed the problems inherent in discharging pulp waste near the shore. If Simpson had been able to meet its original schedule in placing the offshore effluent pipe, this incident wouldn't have happened—and once a new pipe was installed, it should not happen again.

Everyone who knew Joseph well remembers his wry sense of humor. Maybe, on that frustrating day, less than a year into his tenure as executive of the regional board, he smiled and appreciated the irony of the situation. One thing that's certain is that he wouldn't have had much time to dwell on it. Time and pollution moved on.

In the long stretch of northern California for which Joseph's board was responsible, water pollution problems were diverse. The city of Santa Rosa was releasing a growing volume of wastewater into the Russian River, which had gone from a pristine salmon stream to a flowing sewer, with clumps of toilet paper and urban debris on its banks. Smaller cities, including Arcata and Eureka, had ongoing problems with failing sewage treatment systems. The lumber industry was building a sprawling network of logging roads and clear-cuts, which sent loads of dirt tumbling into waterways. The silt destroyed habitat for salmon and trout. Since the California Board of Forestry ignored these impacts, Joseph tackled them.

"Dave started to regulate logging from the standpoint of water pollution," remembers Craig Johnson, an engineer who worked at the regional board in the

1970s and 1980s. "We were telling people to do things above and beyond what the Board of Forestry required."

As had long been the case for the regional water pollution boards, the Board of Forestry was controlled by representatives of the industry it was supposedly meant to regulate. In 1971, a California appellate court ruled that the existing Forest Practice Act was unconstitutional—because it gave power over the environmental impacts of logging to those who profited from the industry.[16] For a year after the court decision, while the state legislature scrambled to build a new forestry law, the lumber industry was essentially unregulated—except on the north coast.

"Dave and his board were the only law in town," remembers his wife, Marion. "For a year, they held the lumber industry at bay based on its pollution of water."

Joseph's job involved a complex mix of science and politics. Regional board members were political appointees, and were known to include ranchers, loggers, or industrialists whom Joseph had cited for violation of water quality standards. At times, his own board seemed as unwilling to act as Shackleton's had been in the bad old days. In 1972, during Ronald Reagan's first term as governor, Joseph let off some steam by writing a sardonic letter to himself signed by an imaginary outraged citizen:

> Why is your board still screwing around, when months ago more than enough evidence was presented to make its responsibility crystal clear? Would you please pass on to your incompetent, impotent and/or self-seeking Board and its senile, Neanderthal chairman my plea that they either get off their butts and do their jobs, or admit to the puzzled world that they are really the lackeys of timber and other industrial interests?

By the mid-1970s, Joseph was taking on the timber industry's common practice of spraying herbicides from small planes. The intent was to knock back the growth of unwanted plants that would compete for sunlight with redwood and Douglas-fir seedlings sprouting in clear-cuts. The label directions on the two most commonly used herbicides, dichlorophenoxyacetic acid (2,4-D) and 2,4,5-trichlorophenoxyacetic acid (2,4,5-T), warned users not to get the spray into waterways.

Industry claimed that they had the spray planes fly on still days, so that the poison never wafted into streams. Monitoring for traces of pesticide was impractical, but they assured the board there was nothing to worry about.

Sampling for chlorophenoxy herbicides was time-consuming and expensive, but Joseph and his staff came up with a quick, affordable way of proving that aerial spraying did leave herbicide in streams. They used fluorescent dyes, a routine means of tracking leaks from sewer pipes.

"The timber companies used a tank to mix herbicides with water," explains Johnson. "We just took a tank and mixed the dye into the water in the same proportion as herbicide, and had a plane spray that. Using a fluorometer, you can detect very small amounts of dye, at levels you can't see." The new technique proved that aerial spraying did indeed put pesticides into streams, tainting drinking water sources and potentially affecting aquatic life. Armed with this new evidence, Joseph required timber companies on the north coast to add fluorescent dye to their herbicide mixtures and to set up instream fluorometers to detect contamination. He set a limit of 10 parts per billion (ppb) on the amount of herbicide allowed in waterways.

This logical response became a political hot potato. If herbicide spraying was curtailed on timber lands, the same thing might happen in the vast farm fields of California's Central Valley. Agribusiness, one of the most powerful players in California's economy, worried that Joseph's board might trigger a movement to limit statewide use of pesticides.

"We were regularly getting calls from the governor's staff, and from the Department of Food and Agriculture," remembers Johnson. "Dave was under tremendous pressure, but as always he stuck to his guns."

Eventually, Joseph banned aerial spraying of phenoxy herbicides within his jurisdiction. The US Environmental Protection Agency (EPA) soon followed suit, banning aerial spraying of 2,4-D and 2,4,5-T nationwide.

In his free time, Joseph relaxed by taking long drives on back roads with his wife, in his cherished BMW. "He knew every bit of his board's territory," Marion says. "All of the North Coast was Dave's natural habitat." He also loved to putter on the small farm in Marin County where he and Marion raised their three kids. He'd rest in the shade of the barn to savor his cigarettes: Marion refused to let him smoke in the house, because she figured the trips to the barn slowed his smoking down. He didn't kick his tobacco habit until near the end of his life, when he was already sick.

In 1985, after serving as executive of the regional board for twenty years, Joseph retired. That same year, 2,4,5-T, one of the pesticides he'd stopped timber companies from dropping out of the sky, was banned in the US. The manufacture of 2,4,5-T creates dioxin, a carcinogenic compound that accumulates in the tissues of fish, wildlife, and people and persists in the environment for years. Every batch of 2,4,5-T is contaminated with dioxin to some extent.

Around the time of Joseph's retirement, a startling link between pulp mills and dioxin pollution was becoming clear. In 1983, EPA scientists found high levels of dioxin contamination in fish downstream from several pulp mills in Wisconsin. This was the first sign that pulp mills produced dioxins; later studies found that toxic organochlorine compounds, including dioxins and furans, were formed during the process of bleaching pulp with elemental chlorine. The agency had not expected this result, and proceeded to hush it up.

Dioxins can suppress immune responses and cause liver damage, rashes, cancer, and reproductive disorders.[17] In 1979, EPA ordered an emergency halt to the use of 2,4,5-T. This decision triggered years of legal and political battles with pesticide manufacturers, led by Dow Chemical.

Starting in 1981, under the Reagan administration, EPA officials began to gag agency scientists studying dioxin contamination and to focus on managing public opinion of dioxin rather than the chemical's health risks.[18] In early 1983, EPA Administrator Anne Gorsuch resigned and her colleague Rita Lavelle was dismissed amid Congressional investigations of their handling of Superfund hazardous waste sites. One of the most controversial examples was the severe dioxin contamination at Times Beach, Missouri, which eventually led to the town's permanent evacuation by order of the Centers for Disease Control. Gorsuch's aide and successor, John Hernandez, also resigned when Congressional hearings revealed that he'd allowed Dow Chemical to edit a report on dioxin contamination caused by its plant in Midland, Michigan.[19]

The link between pulp mills and dioxin pollution first became public when Greenpeace activists published a report called *No Margin of Safety*, in August 1987.[20] In September 1990, EPA issued an advisory against eating fish caught near 20 US pulp mills. Dioxin was accumulating in the tissues of fish, reaching levels that would create a significant risk of cancer and liver damage in people who ate them. EPA's list of the nation's worst pulp mill polluters included the Simpson mill at Fairhaven.[21]

Meanwhile, scientists had begun long-term studies of wild fish living near Canadian pulp mills. One especially revealing example was a mill that discharged effluent into Jackfish Bay on Lake Superior. Researchers found impacts on fish there that paralleled the toxic effects of dioxins on mammals, including humans. The white sucker, a native Great Lakes species, showed damage to its reproductive system, including an increased age to sexual maturity, failure of gonads to develop, and a drop in circulating reproductive hormones. The fish also had enlarged livers, and blood tests showed they were producing cytochrome P450 enzymes, proteins produced by the liver in response to toxic insult. These symptoms of toxicity were much less extreme in fish living near pulp mills that did not use the chlorine bleaching process, which implied that dioxins and furans played a role. When the Jackfish Bay mill was shut down for eight months, all the fishes' symptoms vanished, but some of the reproductive problems returned after the mill started back up with more sophisticated wastewater treatment. Pulp and paper mill effluents can contain over 250 known chemicals: Some of the toxins in pulp mill effluent remain unidentified.[22]

Thirty years after Joseph had first fought to control pollution from the Simpson mill, it and the neighboring Louisiana-Pacific mill were dumping a combined total of 40 million gallons of untreated waste into the Pacific every day. In 1987, the two companies had finagled a waiver from Congress, exempting them from the requirement for secondary wastewater treatment that had by then been imposed on most other polluters dumping into ocean waters off the US coast. The pulp mill outfalls emptied near the North Jetty at the entrance to Humboldt Bay, a popular surfing spot. The waves there turned black from mill waste. Surfers were getting sick, suffering from nausea, sinus infections, headaches, sore throats, and skin rashes. Dioxins and furans were detected in fish and crabs collected near the outfalls, and in the mills' smokestack emissions.

Fed-up Humboldt locals brought their complaints to the Surfrider Foundation, an environmental nonprofit started in the mid-1980s by surfers concerned with the ongoing pollution of California beaches. The group delved into records for the two Samoa mills and discovered that waste discharges routinely violated EPA standards. Surfrider filed suit against the mills in 1989. The EPA, which had for years been trying to get the mills to clean up their effluent, later joined the suit.[23] In September 1991, Simpson and Louisiana-Pacific settled, agreeing to invest more than $50 million in improving their waste treatment. Simpson would use an alternative, chlorine-free bleaching process to prevent formation of dioxins and furans. The mills would be required to regularly test their effluent for toxicity to abalone, sea urchins, sand dollars, and kelp. It was, at the time, the second-largest financial settlement made under the Clean Water Act, and a sweet victory for anyone who cared about protecting the ocean.

"The case proves," said Surfrider attorney Mark Massara, "that the 167 million people who annually use California's beaches can take back our shoreline."

David Joseph would surely have cheered, but by then he was gone. He died of cancer in May 1991, at the age of 64. The people who worked most closely with Joseph remember him as a great man.

"He was fundamental in moving water quality protection forward in California," says Susan Warner, who worked for Joseph at the regional board and eventually succeeded him as executive director. "He was a driven environmentalist. He didn't suffer fools gladly, or tolerate claims unsupported by evidence. Industry saw him as the big heavy."

John Hannum, an engineer who was hired by Joseph to work for the regional board, remembers him with fondness. "Dave was a consummate biologist and a hugely capable motivator," he says. "He knew how fish lived and how water politics played, and he had a healthy skepticism about engineers and their notions. He'd keep us mindful that we were protecting a public resource with public money."

Wes Chesbro was a member of the Arcata city council that battled Joseph in the 1970s. In 2014, when term limits ended his tenure as an elected member of the California Assembly, he noted that his view of Joseph had transformed with time. "I've spent 35 years learning to see things from the other guy's point of view. That's part of being successful in politics," he said. "The regional board was used to dealing with recalcitrant entities that just didn't want to fix things. Arcata was completely outside the box. I have plenty of respect for Joseph's body of work, but we wanted to do something that was untested and experimental. That was too much a challenge to his regulatory regime for him to take it seriously."

NOTES

1. Matthews, G. (June 2, 1957). "F-G chief claims 'hands tied' to stop pollution." *San Francisco Chronicle*.
2. Review, S.L. (1951). "California's water pollution problem." *Stanford Law Review* **3**(4): 649–666.
3. California Legislature. (1957). "Transcript of proceedings, San Pedro, California, October 4, 1957." California Legislative Assembly Interim Committee on Fish and Game.
4. Reish, D.J. (1955). "The relation of polychaetous annelids to harbor pollution." *Public Health Reports* **70**(12): 1168–1174.

5. Rambow, C.A. (1964). "Pollution study of a future tidal estuary." *Journal of the Water Pollution Control Federation* **36**(4): 520–528.
6. California Legislature (1957). "Transcript of proceedings, San Pedro, California, October 4, 1957." California Legislative Assembly Interim Committee on Fish and Game; California Legislature (1958). "Transcript of hearing, Newport Beach, California, July 1–2, 1958." Assembly Interim Committee on Conservation, Planning and Public Works; Subcommittee on Bay and Water Pollution.
7. Reish, D.J. (1955). "The relation of polychaetous annelids to harbor pollution." *Public Health Reports* **70**(12): 1168–1174.
8. Young, P. (1964). "Some effects of sewer effluent on marine life." *California Fish and Game* **50**(1): 33–41.
9. California Legislature (1958). "Transcript of hearing, Newport Beach, California, July 1–2, 1958." Assembly Interim Committee on Conservation, Planning and Public Works; Subcommittee on Bay and Water Pollution.
10. Cloern, J.E., Alan D. Jassby (2012). "Drivers of change in estuarine-coastal ecosystems: discoveries from four decades of study in San Francisco Bay." *Reviews of Geophysics* **50: RG4001**.
11. Reasons, G. (October 16, 1963). "Industrial pollution takes high fish toll." *Los Angeles Times*.
12. Ibid.
13. Kennedy, H. (September 28, 1967). "Water board rejects US advice on purity." *Los Angeles Times*.
14. West, R. (April 11, 1969). "Harbor cleanup plan called 'unreasonable': oilmen rap marine life project." *Los Angeles Times*.
15. Oberrecht, K. "Effects of pulp mill effluents." http://www.oregon.gov/dsl/SSNERR/docs/EFS/EFS14pulpmill.pdf.
16. Lundmark, T. (1975). "Regulation of private logging in California." *Ecology Law Quarterly* **5**(1): 139–188.
17. EPA (1997). "The pulp and paper industry, the pulping process, and pollutant releases to the environment." Fact Sheet.
18. Van Strum, C., Paul Merrell (1987). "No margin of safety: a preliminary report on dioxin pollution and the need for emergency action in the pulp and paper industry." *Greenpeace USA*. http://dioxindorms.com/NoMarginOfSafety.pdf.
19. Shabecoff, P. (March 16, 1983). "Scheuer says EPA aide let Dow delete dioxin tie in draft report." *New York Times*.
20. Anonymous (1989). "Dioxin." *Alkaline Paper Advocate* **2**(2). http://cool.conservation-us.org/byorg/abbey/ap/ap02/ap02-02/ap02-202.html.
21. Savage, D. (September 25, 1990). "Fish taken near 20 paper mills tied to cancer risk." *Los Angeles Times*.
22. Bowron, L. K., K.R. Munkittrick, M.E. McMaster, G. Tetreault, L.M. Hewitt (2009). "Responses of white sucker to 20 years of process and waste treatment changes at a bleached kraft pulp mill, and to shutdown." *Aquatic Toxicology* **95**: 117–132; Lindholm-Lehto, P., Juha Knuutinen, Heidi Ahkola, Sirpa Herve (2015). "Refractory organic pollutants and toxicity in pulp and paper mill wastewaters." *Environmental Science and Pollution Research* **22**: 6473–6499.
23. Paddock, R. (September 10, 1991). "Surfers force pulp mills to halt ocean pollution: suit brings about precedent-setting accord." *Los Angeles Times*.

5

Strangled Waters

First Wave

On a balmy day in June 1955, George Anderson took his sailboat out on Lake Washington, the long stretch of fresh water that separates Seattle from its eastern suburbs. Anderson had recently finished his doctoral research on phytoplankton, and knew the lake well. The water that day looked odd; he noticed a strange brown tinge. So he collected a sample in an empty beer bottle and brought it back to the University of Washington lab where he worked with his mentor, W.T. Edmondson, the ranking authority on the lake.

Under the microscope, Anderson and Edmondson found a life form they'd never seen before. It grew in long, narrow chains, striated with lines that separated one cell from the next. They thought this might be a species infamous among limnologists, the cyanobacterium *Oscillatoria rubescens.* (Cyanobacteria, popularly known as blue-green algae, are in fact distinct from and far more ancient than algae. They appeared more than 3 billion years ago, when the planet was inhabited only by microbes, and were the first organisms to evolve photosynthesis. Their proliferation and release of great volumes of oxygen profoundly changed the chemical makeup of Earth's atmosphere, making the evolution of complex life possible.)[1]

The researchers needed to be sure, so they sent a sample off to an expert, who confirmed their suspicions. *O. rubescens* signaled deteriorating conditions in Lake Washington. To Edmondson, it also meant an unprecedented opportunity to track the impacts of nutrient overload.[2]

O. rubescens had been the harbinger of drastic change in a number of western European lakes. The best-known case was that of Lake Zurich in Switzerland. Fed by Alpine glaciers, Lake Zurich was, until the late 1800s, an expanse of blue known for its abundant populations of whitefish and lake trout, which thrive in deep water. The lake is made up of two basins separated by a narrow passage. In the late nineteenth century towns at the edge of the lower basin, the Untersee, abandoned privies for flush toilets, and began to release their raw sewage into the lake.

In the summer of 1898, a bloom of *O. rubescens* turned the Untersee a bright shade of magenta. Trout and whitefish vanished. The upper basin, the Obersee, received little sewage and remained clean and populated with the normal array of native fish, serving as a control for the unplanned experiment in pollution of the lower basin.[3] By the 1930s, *O. rubescens*, the cyanobacterium that had never been found in Lake Zurich before sewage began to flow, had become dominant. In 1936 it formed a stinking red scum that covered the Untersee's surface.[4]

Lakes scattered throughout North America and Europe were replaying Lake Zurich's disaster. All these bodies of water were choking on an overdose of nutrients unleashed by human actions—a syndrome ecologists named *eutrophication*, based on the Greek phrase for "well fed." In summer, these lakes would turn bright green with dense blooms of algae (or sometimes, as in the case of Lake Zurich, an angry red). As the algal cells died off, they sank to the bottom, where bacteria consumed oxygen in the process of decomposing them. As a result, dead zones appeared—areas of deep water drained of dissolved oxygen, where fish could not survive. Over time, eutrophication has driven the extinction of several species of whitefish endemic to the lakes of the European Alps.[5] It has profoundly altered aquatic ecosystems in North America, too.

Edmondson's first glimpse of *O. rubescens* was "wildly exciting," he wrote.[6] It meant that Lake Washington was in worse shape than he had expected. But it also meant a chance to "turn the clock back fifty-seven years and see what must have happened to Lake Zurich."

Seattle had dumped raw sewage into Lake Washington in the city's early decades. By the 1930s the lake's waters were unsafe for drinking or swimming, so Seattle diverted its sewage to Puget Sound. As the city continued to grow, small towns sprang up on Lake Washington's eastern shore. In 1955, ten secondary treatment plants were pouring 20 million gallons of effluent a day into the lake from the eastern suburbs. Secondary treatment breaks down organic matter, but it doesn't remove nitrogen and phosphorus, which fuel algal blooms. In fact, the treatment process releases these nutrients in inorganic forms—like phosphate (PO_4) and nitrate (NO_3) that are readily absorbed by algae. A 1950 study by Edmondson's students showed that the concentration of phosphate in the lake had doubled, and levels of dissolved oxygen in deep water had begun to decline. Edmondson's lab tracked a steady increase in algal abundance, even before the appearance of *O. rubescens*. The water became progressively murkier. Transparency, measured by the depth to which a white, eight-inch Secchi disc is visible below the surface, decreased from twelve feet in 1950 to two and a half feet in 1958.[7]

On Seattle's western waterfront—in Elliott Bay, an arm of Puget Sound, and along the shore of the Sound itself—water quality was terrible. Four primary treatment plants discharged to Elliott Bay, and forty-eight outfalls dumped 70 million gallons of raw sewage into Puget Sound daily. The Sound's beaches stank, and most were closed due to heavy contamination with coliform bacteria.[8]

A Seattle lawyer, James Ellis, was trying to work out a political answer to greater Seattle's water pollution dilemma. Ellis represented several of the small sewer districts on the eastern shore of Lake Washington. He knew that none of the districts had the resources to handle their growing sewage loads, and believed the only lasting solution would require the separate communities to work together. Engineers at the Bellevue sewage district, in the center of the east shore, came up with the idea of building a sewer line that would extend south, collecting effluents from other towns along the way, and discharge into Puget Sound. The neighboring sewer districts refused to cooperate, so the idea had to be put aside.

In 1956, Edmondson read a newspaper article that implied a committee led by Ellis was considering enlarging the existing secondary sewage plants on Lake Washington. He wrote to Ellis, explaining the basic principles of limnology and the reasons why more treatment wouldn't solve the problem. Ellis replied with enthusiasm. He understood Edmondson's point—the issue in Lake Washington was heavy loads of nitrogen and phosphorus, which would not be controlled by secondary treatment. He agreed that the ultimate solution must be to divert sewage from Lake Washington to the deep waters of Puget Sound, far from shore. This would involve a major public works project, to build a sewer line that would collect wastewater from all the treatment plants on the lakeshore and carry it to the sound.

The first step was to create a government agency that would have jurisdiction over the entire region and the ability to increase sewage bills for every household served. Ellis drafted legislation to create the Municipality of Metropolitan Seattle, called Metro for short. His original proposal would have given Metro control over garbage disposal, water supply, transportation, and parks in addition to sewage. It was rejected by voters in the spring of 1958.

During the hot summer of 1958, the lake began to stink, and many beaches were closed. Mayors of some of the suburban towns took the lead in reviving a simplified version of Metro, with sewage disposal as its only mission.[9] The League of Women voters and hundreds of other volunteers campaigned for the new legislation. To make their point, Metro advocates used a photo of a sad clump of children at a beach where swimming was forbidden. They repeated warnings from Edmondson that unless things changed, Lake Washington would continue to get worse.

The revised Metro was approved by voters in the fall of 1958. The new law enabled the sale of revenue bonds to fund sewer line construction. Sewer bills for Seattle-area homes went up by about $2 per month.

Sewage was diverted from the lake gradually, as construction moved ahead and the interceptor pipe reached more of the lakeshore treatment plants. The first diversion was finished in 1963, removing 28 percent of the total sewage effluent. Lake Washington's phosphorus levels, algal abundance, and turbidity, which had been increasing steadily for years, stabilized. Meanwhile, a new secondary sewage treatment plant was built on the Duwamish River, the lake's outlet, to replace the ten smaller plants to be abandoned on the eastern shore. Three primary plants were built at the edge of Puget Sound, designed to end all raw sewage discharges and move effluents into deeper waters where tides would flush them out to the Pacific. A total of 110 miles of interceptor sewer lines and nineteen pumping plants were needed to channel sewage to treatment plants in the new Metro system.[10]

In 1967, when the new sewer line was nearly complete, water quality in Lake Washington improved dramatically. Phosphorus concentrations and algal densities dropped, and transparency increased to nine feet.[11] By 1971, beaches on the shore of both the lake and Puget Sound were safe and open to swimming.

The rescue of Lake Washington was unique: The people living on its shores had campaigned for the Metro system and voted to pay what it took to clean up the lake. As concern over water pollution grew across the US, the lake's story was trumpeted as a rare model of success. The President's Council on Environmental Quality cited Seattle's Metro project as one of the two outstanding antipollution projects in the nation.[12] (The other example was San Diego Bay, which had also been restored through construction of a vast network of pipes carrying sewage to the sea.)

Ignored was a small sewage rebellion in the town of Santee, California, east of San Diego. When Santee was pressured to sign a forty-year sewage disposal contract with San Diego's regional sewage system, the town refused. "We decided," said Ray Stoyer, head of Santee's water district, "that long before a forty-year disposal contract with San Diego ended, wastewater could become a water resource, not a liability that you pay to get rid of."[13]

Santee already had a secondary sewage treatment plant, which had been discharging into Sycamore Creek. Now the state was requiring that the town end its release of effluent into the creek. Stoyer came up with the idea of filling abandoned gravel mines near the treatment plant with effluent to create a series of lakes. To raise the quality of effluent and ensure the waters would be safe for fishing and boating, he built a series of gravel filtration beds, where treated effluent was made cleaner by bacteria clinging to the rocks. The lakes (Stoyer began with one and ended up with a chain of seven) were stocked with trout and catfish, surrounded with trees and grass, and attracted crowds of anglers and picnicking families, who relished the oasis in the arid landscape. In 1964, after years of extensive testing, the California Department of Public Health announced the fish were safe to eat and the water was safe to swim in.

Southern California relied on expensive water imported from the Colorado River, and later, from the State Water Project, which brought water from the northern half of the state. Santee's reclamation process produced usable water at one-fourth the price of Colorado River water. At one point, the Santee water district predicted it would be able to produce effluent clean enough to pump into drinking water supplies.[14] But by 1974, when Arcata first began to rebel against the state's plan for a regional sewage system in Humboldt County, the popularity of the Santee Lakes had defeated Stoyer's strategy to avoid hooking up to San Diego's regional system.

The lake reclamation system was designed to handle 1 million gallons of wastewater per day, and at first, that was fine. But the greenery and open water created by Santee's sewage drew more than 250,000 visitors a year, and property near the treatment plant and lakes sprouted new homes at a rapid pace. Soon Santee's plant was producing 4 million gallons of effluent per day, the result of increased sewage flow from a growing population. The Santee plant released excess effluent to the bed of the San Diego River, which normally went dry in summer. The sewage flow created year-round pools in the river bed, where mosquitoes bred. The San Diego Health Department ruled that these pools were a health hazard, and that discharges to the river must stop.[15]

The Santee water district was forced to hook up to the Metro sewage system to dispose of its excess sewage. At first town officials feared this would spell the end of the artificial lakes, but they've endured, and remain a popular recreation spot.

San Diego's regional sewage system, meanwhile, was reaching overload. The Point Loma sewage plant, which used only primary treatment, was discharging the sewage generated by nearly 3 million people in the San Diego area to the Pacific. In the late 1980s, the San Diego wastewater system was sued by the US Environmental Protection Agency (EPA) for polluting the ocean, and forced to upgrade its treatment. It was the beginning of the end of the philosophy that the ocean could take whatever humanity dumped in it, that dilution was the solution to pollution.

By 1989, the San Diego Regional Water Quality Control Board, which decades earlier had resisted Santee's reclamation project, was citing it as an example of the kind of creative water reuse that was needed for Southern California's survival. Water demand in the San Diego area was climbing above the amount allotted by the State Water Project, with no new water sources in sight. "It's implicit that we're not going to get more water," said John Foley, chair of the San Diego Regional Water Quality Control Board. "We've got to look at conservation and reuse, or we've got to look at one hell of a change in our life styles."[16] Ultimately, San Diego's Metro sewage treatment system was redesigned to include multiple treatment plants that could channel treated effluent to irrigate parks and golf courses throughout the county—a shift in engineering that would allow sewage to be treated and reused where it was generated, rather than pumping it great distances to dump it in the ocean.

In the mid-twentieth century, though, the idea of elaborate regional sewage projects as the key to saving US waters took hold. It would shape the historic Clean Water Act (CWA) of 1972.

The Great Lakes, five giant bodies of fresh water that fan out along the US–Canadian border, had been presumed immune to eutrophication due to their sheer size. But by 1960s, some of the Great Lakes showed obvious symptoms of decline. On Lake Michigan's southern shoreline, Chicago's beaches stank from the rotting mats of algae that waves deposited on the sand. The city's complicated dodge of routing its sewage away from Lake Michigan and into the Illinois River Valley had only slowed the lake's decline. Chicago's growing population overwhelmed its sewage system, adding to pollution from other lakefront cities and industries.

Erie, the southernmost of the Great Lakes, was known as "America's Dead Sea," the most polluted body of water in the US. An oily scum floated on the surface, along with discarded chunks of wood, beer cans, soda bottles, and the rotting bodies of fish. People were warned away from its beaches because of coliform bacteria in the water, sometimes orders of magnitude higher than levels considered safe for swimming.

Sewage treatment in the cities ringing the lake was primitive. In the Cleveland metropolitan area, only three of fourteen cities had secondary sewage treatment plants in 1966; all discharged their effluents directly to the lake. Cleveland itself had an antiquated sewer system that routinely failed in times of heavy rain. Stormwater overloaded the pipes, causing raw sewage to flow into the lake. David Blaushild, a Chevrolet dealer with a passion for clean water, was suing the city for its failure to stop the discharge of untreated industrial waste into the Cuyahoga River, which bisected Cleveland on its path to Lake Erie. The polluters included Republic Steel, US Steel, Harshaw Chemical, and Reserve Oil Refining Company.[17]

Jeff Reutter, a limnologist who has spent a lifetime studying Lake Erie, remembers swimming there as a kid in the 1950s: Clumps of human feces floated in the water beside belly-up fish. In the 1960s, when Reutter was a teenager commuting to his first job in downtown Cleveland, he passed a rusting metal breakwater on his way to work. It bore a desperate message painted in dripping white letters: "Help me. I'm dying. Signed, Lake Erie."

The most prized commercial fish, walleye and blue pike, were dwindling as Lake Erie's oxygen-depleted dead zone expanded.[18] In the spring of 1965, when Ohio Governor James Rhodes convened a conference on Great Lakes pollution, the situation was dire. Ernest Premetz, chief of the US Bureau of Fisheries based at Ann Arbor, Michigan, informed the conference that blue pike, a species unique to Lake Erie, was extinct. "Walleye are about gone, and whitefish and yellow perch are going," he said. Mayflies, whose larvae live on the lake bottom and form a critical food source for fish, had vanished from Lake Erie after the shallow western basin went anoxic in the late summer of 1953.[19]

The lake was as productive as ever, but because it was overloaded with nutrients, it grew algae and shallow-water fish like carp, sheepshead, and smelt—which nobody wanted to eat. Prized food fish, denizens of deep water like the walleye, became scarcer and smaller year by year. "The commercial catch this year in Lake Erie," noted Premetz, "will be the poorest in history."[20]

Speakers at the conference also discussed the human health hazards of drinking water drawn from Erie and other polluted lakes, the absence or routine failure of effective sewage treatment, and the lack of control on industrial discharges. Within a few weeks of the Cleveland conference, Governor Rhodes sent a letter to the US Department of Health, Education and Welfare pointing out that major sources of Lake Erie pollution lay outside the state of Ohio, and asking that the federal government take responsibility. This triggered an investigation by the US Public Health Service.

Rhodes timed his request to coincide with another water pollution conference, one focused on the plight of the Detroit River, which carried the wastes of Detroit and its industries to Lake Erie. "The Detroit River took on many shades of the rainbow, from pea green, bright lemon-orange, crimson, and deep brown—all streaked with black oil," wrote a reporter for the *Cleveland Plain Dealer*, describing a twenty-mile boat trip the conferees took ending in downtown Detroit.[21]

The Detroit River and Lake Erie had long been important habitats for waterfowl. Antoine de Lamothe Cadillac, Commandant during the founding of Detroit, had described the river's abundance in a 1701 report to his superiors. "The fish there are fed and laved in sparkling and pellucid waters . . . There are such large numbers of swans that the rushes among which they are massed might be taken for lilies," he wrote. The river hosted so much game, one of the native people told Cadillac, "that it only moves aside long enough to allow the boat to pass."[22]

In its lower half, the Detroit River divides into narrow channels braided with shallows. It's an ideal habitat for ducks, geese, and the aquatic plants and small creatures they eat. For centuries after Detroit was settled, the food-rich lower river froze solid in winter, and waterfowl flew away until the thaw. Early Detroit residents raced horses across the ice, and in the Prohibition era, cars loaded with bootleg liquor drove across the river by night. Then, when industrial plants sprouted on the river's west bank in the 1930s, the thaw became permanent. Effluent from auto and steel plants warmed the water enough to keep it ice-free all year, and waterfowl began to stay through the winter. About 50,000 birds wintered on the lower Detroit River in the early years of the auto industry.

Pollution from the plants—especially discharges of oil—would turn the river into a deathtrap. Oil soaks the feathers of water birds, destroying their insulation and leaving them to die of exposure. In warmer weather, oiled birds may slowly starve, unable to fly in search of food. Oil slicks were a regular feature of the Detroit River in the mid-twentieth century. Mass kills of ducks occurred every year from 1948 to 1960, the worst event wiping out an estimated 12,000 birds.[23] Some oiled birds were done in by as little as half a gram of lubricating oil on their feathers.[24]

The Detroit water pollution conference was the perfect moment for Rhodes to make his point. The Detroit River carries 90 percent of the water that enters Lake Erie. In addition to the industrial discharges, Detroit's wastewater treatment plant, among the nation's largest, used only primary treatment and sent heavy loads of pollutants into the river—a trait it shared with plants in many Ohio cities that discharged directly into Lake Erie. "We are past the talking stage, and past the deploring, the study and the study-the-study stage," Rhodes told reporters. "Any further delay is inexcusable. The subject has been talked to death."

The studies and the talking, of course, went on—and so did the pollution. City officials blamed the state. States blamed neighboring states and the feds. People went on passing the buck, flushing their toilets, manufacturing steel and automobiles. Nothing much changed until June 1969, when a spark from a train wheel ignited the Cuyahoga River's chronic oil slick. The heat from that blaze drove a revolution in US water quality law (Figs. 5.1 and 5.2).

Jim Donovan, a longtime environmental activist who worked at the Jones & Laughlin steel mill in the 1960s, describes the blighted Cuyahoga of that era. "It was like a cauldron, with a coat of oil on it and gases bubbling to the surface," he says.[25] "You'd see rats floating downstream, bloated to the size of dogs." Oil slicks floating on the river's surface caught fire on a regular basis, and clouds of black smoke billowed from the water. The problem was an old one: A 1912 river fire had killed five dock workers. The most spectacular river fire blazed in 1952, causing $1.5 million in damage, and might have burned downtown Cleveland to the ground if the wind had not shifted at the right moment.

Figure 5.1 Cuyahoga River fire, 1952. Photo courtesy of the Cleveland Press Collection, Cleveland State University Library.

Figure 5.2 Cleveland Councilman Katalinas, Henry Sinkiewicz, and John Pilch examine cloth soaked in oil from the Cuyahoga River, September 1964. Photo courtesy of the Cleveland Press Collection, Cleveland State University Library.

When Cleveland newspapers reported on earlier river fires, their stories focused on damage to railroad trestles and other industrial structures. The fouling of the Cuyahoga was a routine part of the city's business, seen by many as the normal price of prosperity. By the 1950s, steel mills and other industries had begun to control their releases of oil and grease—not to revive the river, but to protect their property from going up in smoke.

The fire of June 22, 1969, was the Cuyahoga's last. It burned for only half an hour, but it happened at a moment when a national movement for environmental protection was building. Six months earlier, in January 1969, an oil well off the coast of Santa Barbara, California, had blown, spewing 3 million gallons of crude into the Pacific. The tide carried thick deposits of oil and thousands of dying seabirds onto the beaches. The damage was so obvious and intense that hundreds of local people came out to help, trying to soak up oil with bales of straw and attempting to treat the oiled birds. "Never in my long lifetime," wrote Thomas Storke, editor of the *Santa Barbara News Press*, "have I ever seen such an aroused populace at the grassroots level. This oil pollution . . . has united citizens of all political persuasions in a truly nonpartisan cause."[26]

President Nixon traveled to Santa Barbara to view the damage. The federal government carried some responsibility for the disaster: The US Geological Survey had granted Union Oil, the owner of the failed well, a waiver to use a shorter-than-standard pipe casing, and the failure of the casing triggered the well's blowout. In the aftermath of the oil spill, a declaration of environmental rights, written by University of California–Santa Barbara professor Roderick Nash, was read into the Congressional record. Nash's statement, along with widespread grassroots outrage over both the oil spill and the Cuyahoga River fire, framed the conversation that led to passage of the National Environmental Policy Act later in 1969.[27] The new law led to the establishment of the EPA.

Soon after the spill, Senator Edmund Muskie of Maine put forward new legislation that would impose stiff fines for the discharge of oil into US waters. Muskie had plans for more ambitious water quality legislation that would mandate improved treatment of sewage and industrial discharges nationwide. Rather than pushing that broader bill forward right away, however, he decided to wait and see how Richard Nixon's new administration would respond.

In August, *Time* magazine mentioned the river fire in a piece on water pollution. "Some river!" read the article. "Chocolate brown, oily, bubbling with sub-surface gases, it oozes rather than flows." Clevelanders, the article claimed, joked that a man who fell into the Cuyahoga would not drown—he'd decay instead. The story was illustrated with a dramatic photograph of a boat caught up in tall flames on the Cuyahoga—an image captured during the intense 1952 river fire. No photos of the brief blaze of June 22, 1969, are known to exist.

The *Time* story caught the nation's attention. "A river lighting on fire was almost biblical," former Sierra Club president Adam Werbach said in a 1997 interview. "And it energized American action because people understood that that should not be happening."[28]

On June 23, the day after the fire, Cleveland Mayor Carl Stokes held a press conference. He announced that pollution of the Cuyahoga was beyond the city's control, and that he was filing a formal complaint with the state. "We have no jurisdiction over what's dumped in there," he told a reporter for the *Cleveland Plain Dealer*.

State officials argued that even with no industrial discharges at all, the Cuyahoga would remain tainted by Cleveland's failing sewage system. Ben Stefanski, Cleveland's public utilities director, pointed out that any real solution to the problem would take serious money. Cleveland voters had approved a $100 million bond issue to upgrade the city's sewage system in November 1968, but that wasn't enough to do the job. "Without federal funds, we're in a bind," Stefanski said. "We can't tax our people any more . . . and how can we criticize industry for not taking greater steps to halt river and lake pollution when the city itself is one of the violators?"

In September, NBC aired an hour-long television documentary, "Who Killed Lake Erie?" The film was packed with ugly footage of untreated sewage spouting from pipes, scum and debris floating on the lake, dying wildlife, beaches awash in fish carcasses, and grim comments from local people. "One of these days the lake will get thick enough to cross it in a snowmobile," opined the captain of a Niagara Falls tour boat. Leslie Fiedler, a famous literary critic who taught at the State University of New York in Buffalo, described the wind blowing off Lake Erie. "The smell of pollution and corruption that you get is the smell of death," he said. "Not just the death of the lake, but of the people who live by it."

NBC's narrator drove home the point that Lake Erie's plight was just one dramatic example of a nationwide pollution crisis. "Who killed Lake Erie?" his voice intones at the end of the video. "If it dies, we all did. We are all its murderers."

The documentary set off a new round of political finger-pointing. Walter Hickel, Secretary of the Interior in the Nixon administration, announced a series of hearings to look into the failure of Ohio's Water Pollution Control Board to act against polluters. This brought outraged complaints from Ohio officials.

Governor Rhodes laid the blame on Detroit. "We have too many federal bureaucrats who refuse to venture to the source of Lake Erie's pollution," he said. "They ought to go up the Detroit River with a short paddle."[29] Rhodes liked to claim that 90 percent of Lake Erie's pollution came downstream from Detroit—a myth adopted by other besieged Ohio officials. Much of the pollution in the 1960s was discharged directly into the lake from cities and industries in Ohio.[30]

Meanwhile, an alarming realization was dawning on limnologists studying lakes in Europe and North America. The process of over-enrichment with nutrients, algal blooms, and oxygen-depleted waters was speeding up. The acceleration had begun in 1947, when Procter & Gamble put Tide, the first heavy-duty laundry detergent, on grocery store shelves. By 1953, phosphate-based detergents had usurped the market for old-fashioned laundry soap and were enabling the rise of the automatic dishwasher.

Human waste is rich in phosphorus, and sewage effluents had always carried high levels of the nutrient. By1969, detergents accounted for half the phosphates in municipal sewage, and overall phosphorus loads had reached unprecedented levels. Secondary sewage treatment did a good job of breaking down organic matter and reducing bacterial loads but failed to remove elemental nutrients like phosphorus and nitrogen.

In 1969, the same year that *Time* covered the Cuyahoga's final river fire, the US–Canada International Joint Commission released a report on pollution in Lakes Erie and Ontario. The commissioners identified phosphorus as the critical nutrient causing algal blooms in the lakes and urged an immediate decrease in the amount of phosphate used in detergents, and a complete ban within two years.[31] The key role of phosphorus in eutrophication had long been suspected by limnologists, and a federal committee studying Lake Erie had urged drastic reductions in phosphorus loading as early as 1966. But the assumption had always been that the stuff would be removed by adding a chemical precipitation treatment to traditional sewage plants.

The notion that the $45 billion detergent business would have to clean up its own act shocked the industry. At a Capitol Hill hearing called by Representative Henry Reuss, a Democrat from Wisconsin, industry representatives defended their product on the grounds that detergent phosphate was no worse than phosphate in human waste, and that removing the phosphorus from detergents would, on its own, not be enough to halt eutrophication. Reuss, who was sponsoring a legislative ban on phosphate in detergents, pointed out that removing phosphorus from sewage would be an expensive process that could take years to implement. At the time, less than 15 percent of the US population was served by plants capable of secondary treatment, let alone the added step of phosphorus removal.[32] Taking the phosphate out of detergents could cut lake enrichment in half in one fell swoop.

The soap industry took the stance that US citizens would be unable to cope without phosphate detergents, and the Nixon administration agreed. "Elimination of phosphates is desirable in concept but not feasible for implementation at this time," a Nixon staffer testified.[33]

Industry scientists began to claim that phosphorus was not the limiting nutrient affecting algal blooms after all; the culprit was carbon, they said. There was no real evidence for this assertion. It made little sense—carbon is taken from the atmosphere by every kind of green plant during photosynthesis, including algae and aquatic grasses. Elemental phosphorus, on the other hand, originates only in certain kinds of rock, and its availability in lakes and rivers had been drastically increased by human doings, including the nineteenth-century shift from privies to sewage systems, the use of artificial fertilizers, and the rise of phosphate detergents.[34]

Pinpointing the nutrients that cause eutrophication can be difficult, because most lakes have only been studied after years of pollution. "If we are to regard artificial eutrophication as a limnological experiment," noted W.T. Edmondson, the limnologist who'd tracked the deterioration and recovery of Seattle's Lake Washington, "there is the awkward problem of a control. Very few situations have the experimental elegance of the upper and lower basins of Lake Zurich."

Edmondson's data on Lake Washington's recovery was, however, the next-best thing. The lake's algal blooms and turbidity had declined at a rapid rate that correlated with the decreasing concentrations of phosphorus in the water as sewage flows were diverted to Puget Sound. Because secondary sewage effluents are rich in phosphorus, the diversion had made a major impact on phosphorus concentrations, and a relatively minor one on those of nitrogen and carbon. The lake's dramatic recovery reinforced the idea that phosphorus was the key nutrient leading to algal blooms in fresh waters.

At a 1971 symposium on nutrients and eutrophication, a long debate followed Edmondson's presentation on Lake Washington. Frank Derr, a scientist working for the detergent industry, made a convoluted argument that attempted to turn the data inside out. Algae had declined because of the slight decrease in carbon loads, he claimed, and the decrease in algal numbers had somehow lowered phosphate concentrations in the lake.[35] "I'm not persuaded," Edmondson drily responded. His colleague, Joe Shapiro, put it more strongly: "That's like claiming that lung cancer causes cigarettes," he said.[36]

John Vallentyne, a Canadian limnologist with a knack for cutting through the clouds of verbiage surrounding pollution controversies, pointed out a correlation between increasing algal blooms in the Great Lakes and the rising use of phosphate detergents. He charged the US Soap and Detergent Association with a deliberate campaign to confuse the public about the most straightforward way to control eutrophication. Vallentyne was head of the new Eutrophication section at the Freshwater Institute in Manitoba, a part of the Canadian Department of Fisheries and Oceans.

"Limnologists will still be arguing 100 years from now about whether this or that nutrient is more growth-controlling in different lakes," he told the symposium. "Regardless . . . phosphorus can be made to be growth-limiting by removing it from man-made sources that enter the lake."

Vallentyne had already helped to convince the Canadian government to ban phosphate detergents in provinces within the Great Lakes watershed. He was fed up with the US government's failure to act, but he also had an ace in the hole. In the Canadian wilderness, a group of scientists he'd recruited were at work on an unprecedented series of experiments that would silence the loud protests of the detergent industry.

David Schindler spent the winter of 1968–69 dragging heavy bags of synthetic fertilizer across the frozen surface of a series of pristine lakes in remote northwestern Ontario. The lakes lay in a roadless area, so Schindler and a colleague, Greg Brunskill, built some sleds they could load with bags of fertilizer and tow behind snowmobiles.

That year, the lake ice was slushy, making it impossible for the gutless snowmobiles of the time to cross. (The snow machines of the 1960s packed only twelve-horsepower engines. Unlike the modern versions, which zip over snow and ice pushed by more than 100 horsepower, they wallowed and lurched.) So Schindler and Brunskill tramped out five miles of trail in their snowshoes, then returned to run the snowmobiles across the track they'd created. The last half-mile of their journey was a steep, twisting trail full of sharp turns to avoid boulders and trees. The two men had to unhitch their sleds, sling a single eighty-pound bag of fertilizer across the snowmobile's saddle, and sit on top of it while they weaved their way to the untouched lake they intended to pollute.

"We thought it was fun," Schindler would recall forty-six years later. "It was a big adventure."

Schindler, then an ecologist in his late twenties, had been assigned to set up the Experimental Lakes Area. He would head a band of scientists who strategically polluted untouched lakes in order to puzzle out the specific cause of eutrophication. In the autumn of 1968, Schindler set up a makeshift camp in an Ontario forest dominated by black spruce and jack pine. The rocky soil of the region was so thin that the trees took eighty to one hundred years to grow to a height that made them worth the effort to cut them down and haul them to a lumber mill.[37]

There was little organic matter running off the surrounding landscape, so the many small lakes in the region held little carbon. Schindler chose a lake numbered 227, which had the lowest levels of dissolved carbon ever recorded in the annals of limnology, as the ideal place to test the detergent industry's hypothesis that carbon drives the growth of algal blooms. When he and his colleagues added a mix of nitrogen and phosphorus to Lake 227, it was transformed from clear water to a teeming green soup in a matter of weeks. The phosphate they added to the water was instantly taken up by algae, which drew carbon out of the atmosphere to support their growth.[38]

To confirm the finding that algal blooms are created by an overload of phosphorus, Schindler altered another small lake. Dubbed Lake 226, it was shaped like an hourglass, with two basins separated by a narrow neck. Schindler used a vinyl sea curtain to separate the two basins. Then he ran his experiment, adding nitrogen and carbon to both sides of the lake but phosphorus only to one. The phosphorus-laden northeast basin soon produced a bloom of cyanobacteria. Schindler published his results in the journal *Science* in May 1974, effectively shutting down the argument that carbon could fuel eutrophication. An aerial photo of Lake 226, its southwest basin clear while its phosphorus-polluted northeast half held a dense algal bloom darker than the surrounding forest, ran alongside his article. That striking image became an important tool in convincing policymakers in North America and Europe that phosphorus was the critical element to control in fresh waters.

"It was a very convincing example of the old adage that a picture is worth a thousand words," says Schindler. "Often the people making the decisions had no science background, but they understood what that picture meant."

In his *Science* article, Schindler urged the US government to ban phosphate detergents, as the Canadian provinces bordering the Great Lakes and the states of Indiana, Michigan, and New York had already done.

Estuaries around the US and the developed world were also suffering from eutrophication. In Galveston Bay on the Texas Gulf Coast, fish kills caused by oxygen depletion happened on a regular basis. A heavy rain in Houston flushed polluted water into the lower bay in 1968, killing 30,000 fish.[39] The stench of rotting carcasses was intense enough to drive people out of their waterfront homes. Chesapeake Bay was smothering under an onslaught of nutrient-rich runoff from farms and city streets. When Tropical Storm Agnes struck in 1972, the load of silt that flowed downstream smothered aquatic grasses from the mouth of tributary streams like the Susquehanna River to the salty lower bay.

Estuaries and marine coastal ecosystems would prove to be strongly affected by nitrogen pollution. Eutrophication in these habitats would boom during the 1970s due to increasing nitrogen loads from artificial fertilizers and atmospheric fallout of nitrogen from fossil fuel combustion. It would take a long time for scientists to puzzle out the differences between marine and freshwater eutrophication, and even longer for regulators to begin to respond. The EPA began regulating discharges of phosphorus in the 1970s but did not begin a comparable process for nitrogen until 2001.[40]

As the science of water pollution advanced in the 1960s and 1970s, environmental law was also evolving. Legal protection of ecosystems was a new concept, and crafting legislation for the purpose was a struggle. Industrial polluters opposed a strong federal law controlling water pollution. So did the Nixon administration and officials in many of the states.

State and local governments had always held power over decisions on development and discharges in and around rivers, lakes, and estuaries. Waterways were given designated beneficial uses, which often gave a stamp of approval to pollution. Beneficial uses might range from swimming and fishing to, as in the case of the inner harbor in Los Angeles, acting as a cesspool for industrial waste. In the mid-twentieth century, as awareness of water pollution problems grew, Congress made some attempts to force the issue. The Federal Water Pollution Control Act of 1948 had, for the first time, authorized the US Public Health Service to intervene to abate interstate pollution. The federal agency had no real enforcement powers, however, and any action it might take was subject to the approval of state officials. Congress voted to amend the law in 1956, adding a provision for substantial federal grants to aid the construction and upgrade of sewage treatment plants. In vetoing the 1956 legislation, President Eisenhower wrote that water pollution was a "uniquely local blight," and that responsibility for solving the problem should remain with state and local governments.[41]

That attitude changed during the Kennedy and Johnson presidencies. Senator Edmund Muskie, Democrat of Maine, would play a major role in the fight for strong federal antipollution laws. (His counterpart in the House was Representative John Blatnik of Minnesota.) In 1963, as a member of a special Senate Subcommittee on Air and Water Pollution, Muskie introduced a bill that called for creation of a Federal Water Pollution Control Administration (FWPCA). His original bill mandated federal water quality standards for wastewater discharges as well as receiving waters, a radical notion at the time. After a long process of debate and compromise, the bill became law in the autumn of 1965. The final Water Quality Act of 1965 left primary responsibility for water pollution control to the states, which were ordered to set water quality standards for lakes, rivers, and coastal waters—a strategy intended to push states beyond the assumption that waters could be fouled if that was their official use. The FWPCA was created and given limited authority to intervene when the states failed. The idea of regulating pollutant loads in wastewaters had fallen away, but the act doubled the funding of federal grants for wastewater treatment facilities.

By the mid-1960s, the FWPCA had brought abatement actions against automobile factories and other polluting industries on the Detroit River, and against communities whose raw or poorly treated sewage was discharging into Lakes Erie and Michigan. The abatement proceedings moved at a glacial pace through the federal courts. Meanwhile, the federal government was nearly $1 billion behind in funding grants promised to various cities to help improve their sewage plants.[42] The city of Detroit, for instance, had been pledged $10 million in federal wastewater treatment grants, of which it had received only 20 percent.[43]

By 1966, Muskie and his subcommittee had been holding hearings around the country for three years. Again and again, they heard that the major obstacle to water quality improvement was the lack of funding for city sewage treatment plants. Muskie proposed a massive increase in federal construction grants. Congress eventually approved $3.55 billion over five years, a little over half what Muskie had asked for.

In the same year, the US Supreme Court heard a case in which Standard Oil had been charged with violating an obscure law, the Rivers and Harbors Act of 1899, by releasing high-octane aviation fuel to the St. Johns River in Florida. The Rivers and Harbors Act had evolved from a statute written to protect New York Harbor, where in the mid-1800s ships were colliding with heaps of solid waste dumped into the water. An 1886 law prohibited dumping "any ballast, stone, slate, gravel, earth, slack, rubbish, wreck, filth, slabs, edgings, sawdust, slag or cinders, or other refuse . . . into New York Harbor."[44] Over time the statute was broadened to cover waters nationwide. In 1899 the Rivers and Harbors Act was amended to prohibit dumping of "any refuse matter of any kind or description," aside from sewage effluents and stormwater runoff.[45] Any exceptions had to be permitted by the US Army Corps of Engineers.

Standard Oil's attorneys argued that the aviation fuel it had released to the St. Johns was commercially valuable and therefore did not qualify as "refuse."

"Oil is oil," Justice William O. Douglas wrote in the Supreme Court's decision on the case. "Whether useable or not by industrial standards it has the same deleterious effect on waterways." This commonsense ruling came as a bombshell to industrial polluters. The decision made clear that release of any kind of pollutant to US waterways was illegal, unless a permit had been granted by the Army Corps of Engineers. The Rivers and Harbors Act imposed fines on polluters and allowed payment of a portion of these fines to citizens who alerted authorities to violations. The law's resurrection triggered a brief flurry of lawsuits against industrial polluters, brought by individual citizens and environmental groups.

The Court's reinterpretation of the Rivers and Harbors Act also brought forward the notion of a federal agency with the power to control and permit discharges of pollutants. That idea would help build the law that revolutionized water pollution control in the US, the Clean Water Act of 1972. The new law would give regulatory power over city sewage and industrial waste discharges to the EPA, created by President Nixon in 1970.

Nixon objected to everything about the CWA—especially the billions of dollars in federal money it appropriated for sewage plant construction. Nixon was no environmentalist. He created the EPA, a strong step toward federal regulation of pollution nationwide, but only because he was forced into it.[46] Every member he appointed to the National Pollution Control Council, a Commerce Department board, was a corporate executive.

Nixon traveled to Canada to sign the first US–Canada Great Lakes Water Quality agreement in April 1972. At that time he was exerting every kind of pressure he could muster to defeat Muskie's new clean water bill. In November 1971 Muskie's 180-page bill was passed unanimously by the Senate. The legislation, which would become known as the Clean Water Act, contained sweeping changes. For the first time in US history, it declared a legal intent to restore and maintain aquatic ecosystems. It transferred authority over effluents from sewage treatment plants and industrial wastewaters from the states to the federal EPA and set a wildly idealistic goal that all such discharges into US waters should cease by 1985. In a more pragmatic provision, the law required that any city or industry that could not reach the "zero discharge" goal must use the best available technology to treat its effluent.

The notion of technology-based regulation of discharges was unprecedented. Previous federal and state law had required agencies to set allowable levels of pollutants in receiving waters. This practice implicitly acknowledged a right to pollute, and fit with sanitary engineers' traditional philosophy that "dilution is the solution to pollution." The Senate version of the CWA rejected that view.

Under the old regime of setting pollutant standards for receiving waters, any enforcement action required the government to prove that a particular industry or sewage plant was responsible for degrading a body of water. Most often it was impossible to make cause-and-effect links between specific discharges and water quality problems.[47] Enforcement of technology-based standards would be much more straightforward.[48] To enable communities nationwide to build the best available wastewater treatment technology, the Senate recommended an appropriation of $14 billion.

The House was expected to be less amenable to strict new environmental restrictions. The Nixon administration pushed for representatives to rewrite the bill in a way that would restore primary authority to the states, water down federal enforcement, and, most importantly, slash federal funding. Russell Train, chair of the President's Council on Environmental Quality, circulated an estimate that the cost of eliminating all pollutants from industrial and municipal sources by 1981 would run to $94.5 billion. Muskie described this estimate as an attempt "to frighten the people and intimidate the Congress."[49]

Regardless of presidential pressure, the House raised the budget for sewage plant construction from the Senate's $14 billion to more than $20 billion. ("The Nixon administration," says one historian, "was apoplectic."[50]) The main features of the Senate bill survived and the law passed the House with a wide majority. Nixon vetoed it, citing what he called its "staggering" cost. Both the Senate and the House voted to override the president's veto, and the CWA became law on October 18, 1972.

Under the CWA, pollution from sewage and industrial wastes has been dramatically decreased—so much so that even Lake Erie experienced a temporary revival. Still, looking back more than four decades later, major flaws in the legislation, and in the way it was implemented, stand out.

The law failed to effectively address nonpoint source pollution—runoff from farm fields and city streets. In the 1970s, it was already clear that nonpoint pollution was a serious problem. A federally sponsored 1974 study found that urban runoff contributes 40 to 80 percent of the organic matter in affected waters, and the level of toxic material in urban runoff exceeded the amounts in typical industrial discharges.[51] David Schindler predicted in 1976 that the Great Lakes would not recover unless nonpoint pollution was controlled—a forecast that has proven particularly relevant in Lake Erie. Today nonpoint sources are the major cause of pollution in US waters. Runoff is the reason that Chesapeake Bay, Lake Erie, and the northern Gulf of Mexico experience dense algal blooms and contain large dead zones. As for the fate of wetlands, now seen as critical natural filters for nonpoint pollution, the original Act failed to mention the word "wetlands" at all. Still, it has become, at least in theory, a tool in the struggle to keep wetlands from vanishing in the US.

The CWA promised federal money to cover 75 percent of the cost of sewage treatment plant construction, a provision that became one of the greatest public works goldmines in US history. During the first three years of the CWA, President Nixon impounded half the funds authorized by Congress for sewage treatment plant construction: $9 million out of $18 million.[52] Nixon's habit of impounding funds when he disapproved of Congress' spending decisions sparked a running feud over control of the federal purse strings. The issue was resolved by a 1975 Supreme Court decision. In the case *Train v. City of New York*, the court ruled that "the president cannot frustrate the will of Congress by killing a program through impoundment."[53] This led to the release of the CWA funds. Nixon had resigned months earlier.

The delay in funding didn't stop a juggernaut of planning and construction for elaborate regional sewage systems. Engineers were paid a percentage of total costs, giving them an incentive to build the most grandiose projects possible. In the mid-1970s, federal funding for sewage treatment plants was in the billions, approaching the amount spent on hydroelectric dams, flood control projects, and public power plants combined. EPA gave priority to construction of conventional sewage treatment systems, declined to fund projects aimed at treating urban runoff, and set short deadlines for grant funding that forced state and local governments to apply in a rush.

All this intensified the innate tendency of consulting engineers and government bureaucrats to turn to familiar, high-tech conventional treatment. The letter of the CWA required cities to provide secondary treatment by 1983, and the consulting engineers of the nation interpreted this to mean that every town needed an activated sludge system, the approach that cost the most to build and maintain and that sucked down the most energy. More economical kinds of secondary treatment, such as trickling filters and oxidation ponds, were ignored.

As early as 1976, John Rhett, a deputy assistant administrator at EPA, wrote to his superiors to warn them that many small communities were being saddled with elaborate sewage systems they could not afford to build or maintain. He mentioned Humboldt County, California as an example.

EPA would come to understand the problem, but it took time, and resistance from communities scattered around the country. In Mayo, Maryland, a small town on the eastern shore of Chesapeake Bay, locals fought construction of a huge $20 million sewage plant. "Despite their admittedly serious water pollution problems, citizens' groups are calling for smaller, less costly solutions . . . because of the massive development it would permit along the now deserted bay shore," the *New York Times* reported in June 1979.[54] Developers lobbied for major sewage construction projects, while activists in communities scattered across the country began to resist plans for high-tech regional sewage systems. The parallels with Arcata's opposition to the proposed regional sewage system in Humboldt County were obvious. What made Humboldt unique was that its citizens saw the problem coming sooner than most—and that they'd wind up finding their own unorthodox solution.

By 1981, President Reagan was describing the CWA grant program as "a wasteful federal subsidy for urban sprawl."[55] Some environmentalists agreed—a rare event during Reagan's presidency. "We have been spending too much on fancy treatment plants and using the program to help developers spread suburbia," said Larry Silverman, director of the American Clean Water Association. In the early 1980s, federal construction funding was slashed and EPA requirements were changed, allowing communities to keep or construct oxidation ponds and other low-tech treatment systems, so long as effluent quality standards were met.[56]

These changes came after local activists had already waged a long battle against the regional sewage system planned for Humboldt Bay. In their obscure rural corner of California, these passionate sewage objectors had to invent a way to defeat the juggernaut of sewage construction on their own.

NOTES

1. Schirrmeister, B., Muriel Gugger, Philip Donoghue (2015). "Cyanobacteria and the Great Oxidation Event: evidence from genes and fossils." *Palaeontology* **58**(5): 769-785.
2. Edmondson, W.T. (1991). "The uses of ecology: Lake Washington and beyond," p. 13. *University of Washington Press, Seattle.*
3. Hasler, A.D. (1947). "Eutrophication of lakes by domestic drainage." *Ecology* **28**(4): 383–395.
4. Hasler, A.D. (1969). "Cultural eutrophication is reversible." *BioScience* **19**(5): 425–431.
5. Vonlanthen, P., D. Bittner, A.G. Hudson, et al. (2012). "Eutrophication causes speciation reversal in whitefish adaptive radiations." *Nature* **482**: 357–363.
6. Edmondson, W.T. (1991). "The uses of ecology: Lake Washington and beyond," p. 13. *University of Washington Press, Seattle.*

7. Kenworthy, E.W. (1971). "How Seattle cleaned up." *Audubon* **73**(1): 105–106.
8. Ibid.
9. Clark, E. (June 1967). "How Seattle is beating water pollution." *Harper's*, pp. 91–95.
10. Kenworthy, E.W. (1971). "How Seattle cleaned up." *Audubon* **73**(1): 105–106.
11. Edmondson, W.T. (1972). "Nutrients and phytoplankton in Lake Washington." In "Nutrients and eutrophication: the limiting nutrient controversy," pp. 172–193. Proceedings of the Symposium Held at Michigan State University, February 11–12, 1971. G.E. Likens, editor. Lawrence, KS: American Society of Limnology and Oceanography..
12. Kenworthy, E.W. (1971). "How Seattle cleaned up." *Audubon* **73**(1): 105–106.
13. Stevens, L.A. (1974). "Clean water: nature's way to stop pollution." New York: E.P. Dutton & Co.
14. Fradkin, P. (November 12, 1969). "Reused water new source for waning supplies: Santee reclamation plan will bring quality up to drinking standards." *Los Angeles Times.*
15. Stingley, J. (April 25, 1973). "Success may spell death for lakes filled by waste water." *Los Angeles Times.*
16. Bernstein, L. (May 21, 1989). "Water, water (not) everywhere." *Los Angeles Times*, p. J1B.
17. Nussbaum, J. (August 24, 1965). "Citizen foe of pollution wins point." *Cleveland Plain Dealer.*
18. Gale, L. (October 24, 1961). "US biologist warns pollution is destroying marine life in much of Lake Erie." *Los Angeles Times.*
19. Krieger, K. A., Michael T. Bur, Jan J.H. Ciborowski, et al. (2007). "Distribution and abundance of burrowing mayflies (*Hexagenia* spp.) in Lake Erie, 1997–2005." *Journal of Great Lakes Research* **33**(Supplement 1): 20–33.
20. Treon, W. (1965). "The silent waters: lake pollution has reached danger point; control efforts are gathering speed." *Cleveland Plain Dealer*, pp. 1AA–5AA.
21. Treon, W. (June 15, 1965). "US probe of pollution in lake to start at once." *Cleveland Plain Dealer.*
22. Miller, H.J. (1962). "Pollution problems in relation to wildlife losses, particularly waterfowl: Detroit River and Lake Erie." Statement given on behalf of Dept. of Conservation at the Detroit River-Lake Erie Conference on pollution problems, March 27–28, 1962.
23. Ibid.
24. Hunt, G., Howard Ewing (1953). "Industrial pollution and Michigan waterfowl." *Transactions of the Eighteenth North American Wildlife Conference*, pp. 360–368.
25. Cleveland, S.U. (2010). "The Cuyahoga River Fire: don't fall in the river." *Center for Public History and Digital Humanities.* https://www.youtube.com/watch?v=nlHiaZFvcXA.
26. Clarke, K. C., Jeffrey J. Hemphill (2002). "The Santa Barbara oil spill, a retrospective." In "Yearbook of the Association of Pacific Coast Geographers," editor Darrick Danta, **64**(http://www.geog.ucsb.edu/~kclarke/Papers/SBOilSpill1969.pdf), pp. 157–162. *University of Hawaii Press.*
27. Phillips, A. (June 30, 2014). "How a massive oil spill in 1969 changed everything." *thinkprogress.org* (http://thinkprogress.org/climate/2014/06/30/3453277/oil-spill-heard-round-the-world/).

28. Scott, M. (April 12, 2009). "Cuyahoga River fire galvanized clean water and the environment as a public issue." *Cleveland Plain Dealer* (http://blog.cleveland.com/metro/2009/04/cuyahoga_river_fire_galvanized.html).
29. Gale, L. (September 23, 1969). "Rhodes charges Detroit fouls Ohio's Lake Erie." *Cleveland Plain Dealer*.
30. Zimmerman, R. (September 21, 1969). "State, federal government at odds over pollution charges." *Cleveland Plain Dealer*.
31. Schindler, D.W., John R. Vallentyne (2008). "The algal bowl: overfertilization of the world's freshwaters and estuaries," Chapter 7. Edmonton, Alberta, Canada: University of Alberta Press.
32. Ibid., p. 122.
33. Delaney, P. (December 15, 1969). "Detergents held pollution factor: Reuss charges phosphates damage lakes—House hearing starts today." *New York Times*.
34. Edmondson, W.T. (1972). "Nutrients and phytoplankton in Lake Washington." In "Nutrients and eutrophication: the limiting nutrient controversy," pp. 172–193. Proceedings of the Symposium Held at Michigan State University, February 11–12, 1971. G.E. Likens, editor. Lawrence, KS: American Society of Limnology and Oceanography.
35. Schindler, D. W., John R. Vallentyne (2008). "The algal bowl: overfertilization of the world's freshwaters and estuaries," p. 125.
36. Likens, G.E. (1972). "Nutrients and eutrophication: the limiting nutrient controversy," p. 190. Proceedings of the Symposium held at Michigan State University, February 11–12, 1971. G.E. Likens, editor." *W.K. Kellogg Biological Station, Michigan State University* Special Symposia Vol I.
37. McElheny, V. (May 24, 1974). "Carbon cleared in lake pollution: study all but rules it out as key to eutrophication." *New York Times*.
38. Schindler, D.W. (1974). "Eutrophication and recovery in experimental lakes: implications for lake management." *Science* **184**: 897–899.
39. Carter, L.J. (1970). "Galveston Bay: test case of an estuary in crisis." *Science* **167**: 1102–1108.
40. Howarth, R. W., Roxanne Marino (2006). "Nitrogen as the limiting nutrient for eutrophication in coastal marine ecosystems: evolving views over three decades." *Limnology and Oceanography* **51**(1, part 2): 364–376.
41. Hines, N.W. (2012). "History of the 1972 Clean Water Act: the story behond how the 1972 act became the capstone on a decade of extraordinary environmental reform." *University of Iowa Legal Studies Research Paper* **12–12** (http://ssrn.com/abstract=2045069).
42. Hill, G. (April 9, 1972). "Nixon and Trudeau to sign an agreement to fight Great Lakes pollution." *New York Times*.
43. Anonymous (May 27, 1971). "The price of clean water." *New York Times*.
44. Hines, N.W. (2012). "History of the 1972 Clean Water Act: the story behond how the 1972 act became the capstone on a decade of extraordinary environmental reform." *University of Iowa Legal Studies Research Paper* **12–12** (http://ssrn.com/abstract=2045069).
45. Douglas, M.J. (1966). *United States v. Standard Oil Co.* Supreme Court decision. Google Scholar.
46. Ruckelshaus, W. D. (1992). "Oral History Interview." *EPA History Interview-1* https://archive.epa.gov/epa/aboutepa/william-d-ruckelshaus-oral-history-interview.html.

47. Glicksman, R., Matthew Batzel (2010). "Science, politics, law, and the arc of the Clean Water Act: the role of assumptions in the adoption of a pollution control landmark." *Washington University Journal of Law and Policy* **32**: 99–138.
48. Hines, N.W. (2012). "History of the 1972 Clean Water Act: the story behond how the 1972 act became the capstone on a decade of extraordinary environmental reform." *University of Iowa Legal Studies Research Paper* **12–12** (http://ssrn.com/abstract=2045069).
49. Finney, J.W. (November 3, 1971). "Senate approves bill to clean up waterways by '85: votes, 86–0, to shift pollution control authority from states to Washington." *New York Times*.
50. Hines, N.W. (2012). "History of the 1972 Clean Water Act: the story behond how the 1972 act became the capstone on a decade of extraordinary environmental reform." *University of Iowa Legal Studies Research Paper* **12–12** (http://ssrn.com/abstract=2045069).
51. Westman, W. (1977). "Problems in implementing US water quality goals." *American Scientist* **65**: 197–203.
52. Kenworthy, E.W. (February 27, 1974). "G.A.O. says insufficient funds peril deadlines for water pollution control." *New York Times*.
53. Snider, A. (October 18, 2012). "Clean Water Act: vetoes by Eisenhower, Nixon presaged today's partisan divide." *E&E News*.
54. Shabecoff, P. (June 25, 1979). "US clean water program meeting rising resistance." *New York Times*.
55. Shabecoff, P. (July 27, 1981). "Reforms in water plans holding up US funds." *New York Times*, p. A8.
56. Boffey, P.M. (October 12, 1982). "Efforts to gain 'fishable-swimmable' waters appear to falter." *New York Times*.

6

Fighting the Big Sewage Machine

Salmon were George Allen's passion. He spent his professional life seeking to combine salmon restoration with sewage recycling, a mission as daunting as the upstream struggle of a weary chinook blocked by a dam.

Salmon begin their lives as eggs buried in a gravel nest on a stream bottom, from which tiny fish emerge, swim to the surface, and start to feed. The young grow, lose their infant stripes, and swim to sea, steered by instinct and a physical drive to reach salt water. They range through the ocean for two years or more, growing into magnificent creatures. When the time is right, they return to their home streams to spawn. Crowds of wild, abundant salmon once fought their way up the rivers of the Pacific coast from central California to Alaska.

The cycle was eternal, with no distinct beginning or end, until white civilization blocked the rivers with dams and smothered the spawning grounds in silt. By the time Allen came to teach fisheries at Humboldt State, in 1957, salmon runs all along the west coast were depleted. Most of the ancient stocks that had once populated the streams feeding Humboldt Bay were extinct.

In an obscure corner of Arcata's treatment plant, Allen and his students raised young coho and chinook in treated sewage flowing out of the city's oxidation ponds. Soon after he arrived in Humboldt, Allen had begun planning to resurrect the bay's lost salmon stocks, andArcata's sewage oxidation ponds proved the only likely spot to launch his quest (Fig. 6.1). The oxidation ponds were a constant source of fresh water, with access to a stream, Jolly Giant Creek, which formed a small estuary where he could release fish to the bay. He took fingerlings from any hatchery that had extras and raised them to the moment of smoltification, when they lost their baby stripes, turned shiny silver, and transformed from freshwater to saltwater creatures. His intense hope was that they'd go to sea and return as adults, making the city's wastewater plant the center of a salmon revival in Humboldt Bay.

Getting his aquaculture facility built took years—it was first approved in 1963 but, for lack of funding, was not constructed until 1971. Allen persisted. From beneath his bushy eyebrows, he gazed at the world with a singular focus on his fish, often forgetting human details. Three years after one potential grad student wrote to ask if he could study with Allen, he received a reply.

"It was classic George," remembers David Hull, who would go on to become the first manager of Arcata's treatment wetlands. "He wrote, 'Hey, I was cleaning out my office and I found your letter between the wall and my desk. If you're still interested, come on down.' "

Another grad student showed up for an interview and found Allen, having forgotten the appointment, sloshing into his lab coated with muck. He'd been crawling through culverts on Jolly Giant Creek to make sure there were no barriers to any salmon that might return and push upstream to spawn.

Working at the oxidation ponds caused Allen to sprout a second obsession: recycling the nutrients in sewage to grow fish. In the 1960s he spent a sabbatical year in Europe studying wastewater aquaculture projects in Switzerland and Germany. Munich had a system that raised carp and tench in treated sewage. The fish grew fast, feeding on plankton that thrived in the nutrient-rich water, and were sold at a profit.[1] The use of human waste to fertilize fish ponds was, and is, common in Asia. The process works well with carp and tilapia, which feed on detritus and are adapted to murky waters. Salmon need clear, cold water, and Allen was the only fisheries scientist in the US, and likely in the world, trying to raise them in treated sewage.

The process was frustrating. Oxidation pond water carries high levels of ammonia nitrogen. In alkaline conditions, ammonia becomes intensely toxic to fish; in the early years, many thousands of fingerling salmon died of ammonia poisoning. Later, when Allen worked out the quirks and succeeded in raising fish to the smolt stage, he released them near the mouth of Jolly Giant Creek. Herons and egrets lined up on the banks like guests at a buffet, devouring many of the smolts before they could reach the bay.

Conventional salmon hatcheries had the funding and equipment to control water quality and to fence out predators, but Allen had to scrape by on a shoestring budget. "We do things El Cheapo around here," he said. To form substrate that could support the invertebrate creatures he hoped would flourish and feed his salmon, he scrounged broken concrete blocks and bundles of brush cut by the city landscaping crew, which were then submerged in the fish ponds.

No difficulty fazed him. Well before the days of the Clean Water Act and the nationwide push for improved sewage treatment, Allen had impressed Arcata's city government with the idea that sewage was a resource, not just a problem to be flushed away. So it was natural that when a city staffer went looking for an escape from the state-mandated Humboldt Bay Wastewater Authority (HBWA) project, he turned to Allen.

Frank Klopp, Arcata's genial, cigar-chomping director of public works, had let the initial controversy over the regional system pass him by. He was not an environmentalist or an activist of any stripe. If state officials said centralized sewage was the way to go, that was okay with him. On an autumn afternoon in 1976, Klopp was reading through an engineering report on Arcata's sewer system when he noticed a financial projection in the back pages: Once the HBWA project was up and running, Arcata would need to double its sewer rates to pay for the system's upkeep.

Figure 6.1 George Allen at Arcata's wastewater oxidation ponds on Humboldt Bay, 1971. Photo by Peter Palmquist, courtesy Humboldt Collection, Humboldt State University Library.

"I'd just completed a rate increase for the city, in anticipation of the regional plant being built," Klopp remembers. "The last pages of this study said that in two years we were going to have to double our sewer rates *again*."

He winced at the prospect. "As public works director, when you stand in front of the citizens of Arcata and announce a raise in the sewer rates," he says wryly, "you're not universally applauded."

Klopp got up and walked into the city manager's office. Dan Hauser and another city council member, Sam Pennisi, were there. Klopp explained his new discovery about the cost of the regional system: Arcata's portion of the cost of maintenance and operation had risen to $475,000 a year, every year, forever.

"Want to try one more time to get out?" he asked.

"Go ahead and try, if you want," replied Hauser.

Klopp wasn't driven by worries over urban sprawl or the risk of pollution if there was a break in the miles of pipeline planned to carry sewage around the bay and under its floor. "The reason I got into it," he says, "was straight economics. The state water quality board came up with what you get a lot of with government, a one size fits all plan. [The Bays and Estuaries policy] was not taking the world as it is right now and asking how can we protect existing uses and make it better. It was a policy to say, basically, we're going to go for deep water discharges."

To justify Arcata's escape from the HBWA project, Klopp turned to Allen's aquaculture experiment. He proudly called this notion his "brainfart." The state wouldn't allow any sewage effluent to flow to the bay unless it could be proven to enhance the waters. Klopp was an avid fisherman and duck hunter: If using treated sewage to grow salmon and release them to the bay wasn't enhancement, he couldn't imagine what was.

John Hannum, the regional board's engineer, was not impressed. He warned that "Things would go rather hard for Arcata" if it withdrew from HBWA. He dismissed the notion that sewage aquaculture could release the city from the regional system. "Allen managed to get fish to survive in wastewater, but that's not aquaculture," he told a reporter. Hannum knew the aquaculture experiment had suffered episodes of mass ammonia poisoning. Smolts had been released, but at the time of Klopp's brainfart, no adult salmon had yet returned to Arcata. Klopp cheerfully acknowledged that the city was unlikely to ever turn a profit from fish farming. That didn't matter, he said, because "all we're really looking for is an alternative to pumping sewage into the ocean."

Allen had known David Joseph for years and had supported Joseph's fight to stop beach discharge of pulp mill wastes in the 1960s. When they encountered each other again during the long fight for an alternative to the HBWA project, Allen's eccentricities drove Joseph nuts. Between technical arguments, Joseph, a tidy man, would plead with Allen to put on two socks of the same color.

Under Allen's influence, Arcata's crew of sewage activists would eventually come up with a treatment alternative so useful that it has now been imitated around the world. The process took time, however, and it never would have happened if the HBWA project went ahead on schedule. A new pocket of resistance cropped up from an unexpected quarter, led by a soft-spoken peacenik who used the courts to stall the sewage construction juggernaut.

At first, Dan Ihara wanted to make sure that the HBWA project would include his town. He was a member of the Community Services District in Manila, a village of a few hundred people on Humboldt Bay's western shore. Manila had no sewer system, and Ihara figured the community couldn't afford to miss out on the chance for low-cost wastewater treatment being touted by HBWA's engineers.

That changed in September 1976, when he came across some cost projections from a Winzler and Kelly engineer that showed Manila would face much higher sewage costs as part of the regional system than it would in building its own treatment plant. Manila had no official voice on the HBWA board, but it seemed clear the town would be bankrupted by the costs of the regional project if it went ahead.

Ihara, a conscientious objector to the Vietnam War, had spent the early 1970s doing alternative service in San Francisco, scraping by on a stipend of $50 a month. He and his wife moved north to Humboldt to get away from the big city, and landed in Manila where housing was cheap. By 1976, their daughter was a toddler. Ihara was working at Redwoods United, a community development agency, where he helped create and run job programs for disabled adults and disadvantaged teens. He was doing good work for low pay—a pattern that would hold throughout his life.

A mutual friend connected Ihara with Jacqueline Kasun, an economics professor at Humboldt State, who was also worried about HBWA. Kasun lived among the pastures south of Arcata and dreaded the development a sewer line would bring to the area. She had previously fought freeway projects proposed for Arcata and Eureka. "I was identified as an environmentalist," she said. "But I didn't fight those freeways so much on environmental grounds. I'm an economist, and I don't like boondoggles."[2]

In Kasun's eyes, the regional sewage project was wildly overdesigned, with a capacity more appropriate to an urban zone than to mostly rural Humboldt County. One extreme example was the plan for the hamlet of Indianola, where Kasun lived. Homes there were scattered and population was low, less than one person to the acre. "The costs of extending sewer lines . . . to such sparsely settled areas are too high to be practical," she wrote. "This has always been common knowledge, at least until sanitary engineers got into the enormously profitable business of hustling sewer grants."[3] The majority of the HBWA project's massive cost, she pointed out, was for moving sewage as opposed to treating it.

HBWA was about to issue $12 million in bonds to finance the local share of costs for the regional sewer system. Ihara and Kasun consulted with a lawyer, who told them that the only way to stop the project was to give local people a chance to vote against it. They could petition for a referendum on the sewer bonds, but time was short: The bonds would be issued in a few days, unless opponents could gather at least three thousand voter signatures before the October 7 deadline.

They called a community meeting at the Presbyterian Church, a rambling clapboard building in Arcata. On the evening of September 31, about thirty people gathered there. "We asked if there was enough interest to do it," Ihara remembered. "We passed the hat around, and people said, let's go for it. By the time of that meeting, there were seven days left." The newborn Committee for a Sewer Referendum managed to gather more than three thousand signatures and rushed them to the county clerk's office just before the 5 p.m. closing time on the day of the deadline.

The next day, HBWA's lawyer, John Stokes, announced that the petitions had been delivered to the wrong office. Stokes advised the HBWA board that their secretary was the only person who could legally accept the petitions—and that he could no longer do so, since the October 7 deadline had passed. Dan Hauser spoke up. "The filing was made in good faith by the petitioners," he said. "I think we have a strong moral obligation to these people, and I move we accept the petition."[4] No one on the board would second Hauser's motion. Still, Hauser's support marked the beginning of a close alliance between Arcata's sewage rebels and Committee for a Sewer Referendum.

Ihara filed a lawsuit in a bid to force the agency to accept the petitions. Meanwhile, he didn't sit and wait for the legal wheels to grind: He wrote a letter to the federal Economic Development Administration, threatening further litigation if the agency approved a $5 million grant to HBWA. "We urgently request that EDA not authorize a grant which could be used to avoid resolution of issues of public concern and which would undermine the right to petition and call for a referendum," Ihara wrote. John Stratford, general manager of HBWA, said he was "appalled" to hear of the letter. He described Ihara as a "saboteur."[5]

Testimony in court revealed that everyone involved, including HBWA staff, had assumed the county clerk's office was the right place to submit the petitions. "It was like a scene from Perry Mason," Ihara recalled. "They had the secretary of HBWA on the stand, and they asked him where he was at 5:00 on October 7. He was in the county clerk's office, looking out the window for us, wanting to see that we met the deadline. If we had gone to his office at that time, he wouldn't have been there, because he didn't know he was the one who was supposed to receive the petitions."

The judge ruled that Committee for a Sewer Referendum's petitions should be accepted. HBWA appealed the decision. When that tactic failed, HBWA rejected the petitions on the grounds that some of the signatures were invalid—an alleged percentage of the total that varied as the fight wound through the courts.

Ihara and Kasun managed to get hold of rate projections for individual homes and businesses under the HBWA plan. The startlingly high figures won the Committee influential allies, including Lawrence Lazio, head of the local company Lazio Fisheries and a cousin of former San Francisco mayor Joseph Alioto.[6] A number of local business people began to oppose the regional system, including Don Quinn, of Eureka's Coca-Cola bottling plant. "Not only are our sewer rates going to be too high," he wisecracked, "7-Up's are too low."

Chuck Goodwin, chairman of the HBWA board, warned that delay was far too expensive. Inflation on the project was close to $400,000 a month.

David Joseph and his staff made it clear that Allen's aquaculture experiment, even in an expanded version, would never fly as an official enhancement to Humboldt Bay. Hannum, the regional board engineer, was dismissive of Allen's project, viewing it as a preordained failure. "They killed millions of little fish at George's ponds," he said later. "The pH of the water would go up, and then ammonia would dissociate, and ammonia is acutely toxic to little fish. The birds would kind of burp, you know. You never found too many corpses with those frog-stabbers up there."

The inspiration for a marsh came out of long discussions between Allen and Robert Gearheart, an environmental engineering professor who'd come to Humboldt State in 1975. Gearheart had worked with low-tech biological sewage treatment systems in Arkansas, Texas, and Utah. The regional board's insistence on a centralized plant using activated sludge treatment struck him as seriously misguided. The energy-intensive process relies on wastewater being held in the treatment plant for a number of hours, during which a cultivated broth of aerobic bacteria breaks down pollutants. During Humboldt County's winter rainstorms, the system would be overwhelmed with a high volume of sewage, resulting in the release of untreated effluent to the Pacific. These practical problems were outweighed in the eyes of the state's engineers by the fact that activated sludge was in vogue at the time, as were regional systems that pumped sewage long distances to a centralized treatment plant.

"My profession got things going the wrong way," Gearheart remembers. Engineers focused on the large amounts of cash available for construction grants rather than exploring the most efficient alternatives. "The construction grant money was a cash cow. In California in the 1970s, you could get 87 percent of your project paid for by a combination of federal and state money. Engineers were paid a percentage of construction costs, and had no incentive to design less costly systems."

Gearheart doesn't remember exactly when the idea of building marshes as a combination of wildlife habitat and enhanced sewage treatment first dawned. It may have happened one day while Gearheart and Klopp were hanging out with Allen at his aquaculture pond. They looked out over the barren landscape of Mt. Trashmore, the old municipal landfill that had been condemned and capped with bay mud. Salt continued to leach out of the mud, so that nothing grew there. Beyond the gray, lumpen face of Mt. Trashmore lay a heap of decaying logs, which had been piled there years earlier, before one of the nearby lumber mills had shut down. They began to imagine what could be made out of this grim landscape: tall stems of bulrush could rise from marsh ponds that would replace the heap of rotting logs and the weedy field beyond it. Reviving this wasteland could be their safe passage out of the HBWA morass.

Soon after the brainstorming session at the fish ponds, Klopp learned that the Mountain View Sanitary District, on the eastern edge of San Francisco Bay, had constructed a small wetland fed with treated sewage. The wetland was a strategy to avoid the expensive discharge requirements imposed by the San Francisco Regional Water Quality Control Board.

Klopp drove down to Mountain View along with Hauser, Allen, and Gearheart. At the Mountain View Sanitary District, beyond the rhythmic noise of the trickling filter treatment system, where giant rotating sprinklers spread sewage over beds of piled rock, they found a quiet pond. Tucked between oil refineries, the wetland, which had been dubbed Moorhen Marsh, was ringed with bulrush and cattail and full of red-winged blackbirds and marsh wrens. Ducks and coots dabbled in the water. Though the air was tainted with the stink of the refineries, the sewage-fed marsh had formed a green oasis in the industrial landscape.

Discharges of treated sewage were allowed in San Francisco Bay, but the regional board required a dilution of 10:1 at outfall pipes. The only way to accomplish this was with a deep-water outfall, a prospect that was prohibitively expensive for the small Mountain View Sanitary District. The district would have had to build a pipeline more than two miles long to carry its discharge deep into the bay.

The San Francisco Area Basin Plan allowed exceptions to the 10:1 dilution rule if sewage discharge could be shown to create an "environmental benefit," though that phrase was not defined. Warren and Ed Nute, father-and-son consulting engineers working for the Mountain View Sanitary District, came up with the idea of using effluent to create a wetland.

"We decided to see what would happen if effluent was put into a marsh," says Ed. "The district hired a guy with a bulldozer, to dig ponds next to the treatment plant." Treated sewage flowed in, cattail and bulrush sprouted up, and birds appeared. Members of the local Audubon Society chapter began birding at the wetland and reporting rare species there.

The Mountain View Sanitary District argued that the wetland constituted an environmental benefit. "My dad testified at a lot of meetings," remembers Ed Nute. "Regional board staff had the entrenched idea that everything should go through a deepwater outfall and be diluted. Dilution is the solution to pollution, that was their philosophy."

Despite the mass loss of wetlands in the region, and the heavy use of Moorhen Marsh by native birds, it took years to bring the San Francisco regional board around to the idea that constructed wetlands were a valid environmental benefit. Eventually, the board developed a formal marsh policy, requiring a minimum acreage of wetlands to be developed per million gallons of effluent discharged per day. To comply, the Mountain View Sanitary District created a second, much larger wetland, known as McNabney Marsh, now home to a large flock of white pelicans and many other species.

At the time of the Arcata contingent's visit, however, Moorhen Marsh was only a small, appealing experiment. Klopp and his companions drove home, with thoughts of their own conundrum and the peace and birdsong of the sewage marsh percolating in their minds. Gearheart understood that wetlands could improve on established lagoon and oxidation pond systems. As they grew, cattail and bulrush would absorb the nutrient load in treated sewage, while their root systems hosted microbes that would decompose organic matter. A marsh made excellent sense so far as water quality went. Moorhen Marsh had shown the Arcata group an example of a constructed wetland as a political tactic.

Gearheart, Allen, and Klopp put together a proposal for an alternative wastewater treatment scheme that would reclaim Arcata's treated sewage by creating marsh habitat and expanding Allen's sewage-fueled salmon aquaculture. Breaking away from the HBWA project would be a risky endeavor, warned John Corbett, the city's attorney. Resisting the state-mandated plan was a high-stakes poker game because it meant risking the loss of millions of dollars in state and federal sewage construction funding. "If Arcata lost, they would have been bankrupt," says Corbett, who is now a member of the regional board. "The city council members who led the fight—Hauser, Chesbro and Fairless—displayed the kind of raw courage rarely seen in local government. I haven't seen the like of it before or since."

The city council endorsed the proposal in February 1977, and then, as Hauser later recalled, "all hell broke loose."[7] Joseph had his staff deploy every possible bureaucratic tactic to block consideration of Arcata's new alternative. Alex Fairless, Arcata's mayor, wrote to Joseph on April 14, asking the regional board to consider the city's new plan at its meeting on May 26. She asked that the board reinterpret the Bays and Estuaries policy to accommodate the unconventional arrangement envisioned by Gearheart, in which constructed marshes would "polish" treated wastewater flowing from the oxidation ponds, raising it to a higher water quality standard before it reached the bay. She also requested a modification of the Basin Plan to allow Arcata to bow out of the regional system, and a recommendation that the city's project be funded under the federal Clean Water Grant Program.

Joseph wrote back immediately, dismissing the proposal as failing to demonstrate "enhancement" under the Bays and Estuaries policy and declining to give Arcata a spot on the agenda at the board's next meeting. "We sincerely urge that you renew your commitment to the local HBWA project," he wrote. Fairless fired back, insisting that the city be given a hearing in May. "Based on the information proposed in your draft report," Joseph answered, "we believe that whether your waste goes directly to the Bay or traverses a marsh composed primarily of effluent, it still constitutes waste and is still governed by the provisions of the Water Quality Control Policy for Enclosed Bays and Estuaries of California."

Hauser, Gearheart, Klopp, and Allen traveled to the Mendocino County town of Ft. Bragg and sat through the regional board's day-long April meeting before they were given a chance to plead for a hearing. "I was trying to explain the idea of reclaiming wastewater to feed a marsh," Hauser recalled. "John Hannum said, 'What are you going to do? Send it [Arcata's sewage] up in a balloon?' I answered that the purpose of a hearing was to give us a chance to explain."

Hannum and the other staffers did not budge. The members of the regional board, however, were interested citizens, rather than career bureaucrats, and some of them had great sympathy for Arcata's cause. One board member, Andrea Tuttle, was a young biologist living in Arcata, who in the past had argued against the regional sewage system because it would bring on urban sprawl. She was also concerned the system would create a real risk of intense pollution in the event of a break in the planned trans-bay sewage pipeline. Urged on by Tuttle, board members agreed to place consideration of a public hearing on Arcata's proposal on the agenda for the board's May 26 meeting.

At the May meeting in Rohnert Park, regional board staff advised against granting Arcata a public hearing. "If a hearing were allowed, Arcata would have to prove that they would enhance the bay by the development," said Ben Kor, supervising engineer for the regional board. "They have indicated it's not possible to do that." Kor quoted the board's legal counsel to make his circular argument: Enhancement had not been proven, therefore there was no reason for a hearing on the potential for enhancement.

The whole point of a hearing, Arcata's attorney answered, was to have the board itself decide whether the city could enhance the bay with its alternative plan. "The board staff has been against having a hearing whether it's enhancement or not," he said. "All we've been trying to do for two months now is get a hearing."

Board members rejected staff advice and granted Arcata a public hearing at the next scheduled meeting in late June. "We get to show our proof of what enhancement is on our terms," Hauser told a reporter. "We don't have to show it on their staff's terms."[8] The Arcata city council, he said, believed raising fish, creating wildlife habitat, and discharging nutrients in treated wastewater would enhance the bay. He didn't have high hopes for immediate approval, but saw the hearing as a first step in the long process of fighting for their idea.

A week before the hearing, regional board staff announced that Arcata's proposal had to be recast as a formal "facilities plan," an engineering document that normally took months to prepare. Everyone involved in Arcata's sewage fight went into overdrive. "George Allen and Bob Gearheart were dictating, the city staff was typing, Alex Fairless and I were up all night," remembered Hauser. Fairless worked the photocopier in the police department, in the city hall basement, while Hauser copied and collated upstairs. Their passion for the cause had infected the clerical workers. "The secretaries were up all night with us," says Fairless, "and one of them was very pregnant. They'd find a mistake in one paragraph, and then they'd have to retype the whole page." They got the required number of copies of the facilities plan on a bus to Santa Rosa at 7 a.m., then went their separate ways in search of sleep.

The document produced by Arcata's heroic group effort failed to inspire the regional board staff. "On June 16, 1977," a staff memo reported, "staff received a 2.75 pound report entitled *City of Arcata Facility Plan and Project Report, June 1977*. Despite its bulk, the report contains very few passages which staff considers relevant to the issues to be considered at the hearing." The plan, the memo noted, contained long sections plagiarized from Arcata's General Plan and the Metcalf & Eddy report. So far as board staff were concerned, the hefty document failed to show "that water quality would be enhanced above that which would occur in the absence of discharge."[9]

Meanwhile, representatives of the other communities involved in the HBWA project resented what they saw as Arcata's effort to escape a commitment, leaving them to carry a multimillion-dollar burden. Ed Estes of McKinleyville found Hauser and the Arcata contingent condescending and irresponsible. He pointed out that his town had been pushed to make a huge investment: $7.6 million had been spent on installing sewer pipes that still led nowhere, while McKinleyville remained under a building moratorium imposed by the regional board.

"We in McKinleyville believe in contracts and we honor them," he said, with clear disdain for Arcata's last-minute attempt to escape the HBWA agreement. "Private citizens in my community may be, as others have suggested, brush Oakies. We are mostly upper lower class, we are not horribly overeducated. But we are beginning to realize what our individual costs are from this delay."

The Humboldt County Board of Supervisors had opposed the Bays and Estuaries policy before it was adopted and had gone so far as to initiate a lawsuit against the state board, trying to stop its implementation. The county had paid over $1 million to build sewer lines from outlying communities south of Eureka leading to a trans-bay pipeline that did not yet exist. "The county is extremely apprehensive of state and federal declarations that this regional project will resolve our wastewater problems for many years into the future," said Supervisor Erv Renner. Given the track record, Renner feared that the board might suddenly decree that discharging secondary-treated sewage to the ocean was forbidden, after pushing through the HBWA system at vast expense. But if the county and other cities on the bay were to be forced to go regional, he insisted that Arcata should have to do the same. If Arcata found an out, the county would also demand the right to jump ship.

The day of the hearing was sweltering hot at the Ukiah Fairgrounds. "It was 103 degrees in those metal buildings," remembers Hauser. The proceedings ran for more than six hours. Four decades later, the tension of that long, overheated meeting can still be heard through the sound distortion on the warped reel-to-reel audio tape recorded by the board staff. There's a level of emotion—resentment, hope, anger, and sorrow—startling in a bureaucratic hearing on sewage.

Joseph began by summing up his arguments against Arcata's proposal. The city hadn't proved that discharging effluent into a series of marshes that emptied into the bay was substantially different from dumping into the bay itself. His most crucial point he saved for last: If Arcata won release from the HBWA project, the other cities and Humboldt County would demand the same privilege. "The end result will be a continued discharge of waste into Humboldt Bay, which we have been trying to alleviate for the past ten years," he said. Joseph's dream of ending flows of sewage into the state's largest biologically intact estuary would die.

Dan Hauser explained why Arcata was trying to bolt from the HBWA project when it was already so far along—ready to build, as soon as the legal wrangling with Committee for a Sewer Referendum might be resolved. Construction and operating costs were astronomical and had not been fully revealed until a few months before; the project would suck down more energy than any other facility in the county, much of it for moving sewage long distances; and the regional project clashed with provisions in the revised Clean Water Act of 1977, which endorsed water reclamation and energy conservation.

He added that Arcatans were quite serious about addressing sewage problems, as demonstrated when voters passed a recent bond issue to replace leaking sewage interceptors by a margin of 86 percent. "That's an unheard-of majority today for the passage of a bond, and unlike our counterpart, the regional authority, we were not reluctant to submit this issue to the voters."

Hauser's most eloquent argument was an amateur video of islands in the brackish lake next to Mt. Trashmore, where thousands of shorebirds roosted at high tide. A Humboldt State student had climbed a utility pole and caught images of a peregrine falcon hunting over an island crowded with sandpipers. The shorebirds lifted into a hypnotic whirl of synchronized flight as they evaded the predator. The video also showed a white-tailed kite hovering as it searched for prey. The kite, a wetland-adapted raptor, was then just beginning to recover from a brush with extinction in northern California. Building marshes to treat the city's wastewater would expand important habitat, critical to wildlife.

Gearheart explained the workings of natural sewage treatment both in the existing oxidation ponds and in the marshes Arcata hoped to create. The great advantage of this system, he emphasized, was time. The longer wastewater stayed in the ponds and marshes, the cleaner it would get. In the activated sludge process the regional board envisioned, wastewater would be retained for a matter of hours, and heavy pulses of rain could overwhelm the system, causing the release of raw sewage. A drop of wastewater would take weeks to move through the oxidation ponds and marshes Arcata proposed, and by the end of this journey, pathogens would have died off and pollutants would have been broken down.

"The natural function of a marsh is to exist in places where you have high levels of nutrients," Gearheart assured the board. "The ability of marshes to remove BOD, ammonia, nitrate and phosphate is very high." Marshes form in deltas and estuaries laden with nutrient-rich silt that runs off the land. Cattail and bulrush slow the flow of water, causing solids to settle out. Among their roots live a community of microbes that digest organic matter, while the plants absorb nitrogen and phosphorus. The combined oxidation pond and wetland system could be maintained for $300,000 less per year than Arcata's share of operating costs for the proposed HBWA project.

In Joseph's eyes, the marsh argument was untested bunk. He seemed skeptical of the notion that marshes act as natural filters for sediment and nutrients; he doubted that humans could build a working wetland. Gearheart could claim no previous experience in marsh creation.

When George Allen rose to speak about his decade-long experiment with raising salmon in wastewater, he gave too much history; his talk came off as a litany of frustrated hopes. The first handful of adult salmon raised to smolthood in his ponds had returned only that year. Allen saw this as great progress, but Hannum couldn't resist taunting him about the low return rates. A biologist with the Department of Fish and Game gave his devastating take on Allen's aquaculture experiment. Summer water temperatures in the Arcata fish ponds often rose high enough to kill young salmon, if they survived exposure to high concentrations of ammonia. He saw discouraging parallels with an attempt to create a trout fishery at the Indian Lake Reservoir, built to receive treated sewage effluent flowing from developments around Lake Tahoe in the Sierras. An ammonia-stripping device had been built there, but in winter it was too cold for the thing to function, and the reservoir saw repeated mass die-offs of trout from ammonia poisoning.

Aside from a representative of the Redwood Region Audubon Society and Arcata's people, everyone who spoke assumed that regional sewage systems were by nature more reliable and efficient than small local ones, and that they were inevitable: because the engineers had said so, because the state board was pushing for them, because the Bays and Estuaries policy enshrined them. That notion was repeated over the sweaty hours in the hearing room, until a young, shaggy-haired man walked to the podium and unleashed an eloquent critique of the whole notion of regional sewage systems in general, and the HBWA project in particular.

Wade Rose was representing the Governor's Office of Appropriate Technology (OAT), a new niche in the state bureaucracy created by then-governor Jerry Brown. OAT's mission was to research and promote low-energy, ecologically sound technologies. Its director was Sim Van der Ryn, a devotee of composting toilets who wrote a short, punchy book entitled *The Toilet Papers*, in which he explained that regional sewage systems separated humanity from nature in dangerous ways. It was imperative, Van der Ryn claimed, that everyone start "taking responsibility for our shit," through radical, local means. He described pausing during an agonized debate over the permissibility of composting toilets to feed potted plants in a state office building with odorless humus from his home composting toilet.[10]

"It is the opinion of this office," Rose announced, "that the energy crisis of 1972 has left the HBWA proposal as it now stands over-planned, prohibitively expensive, and obsolete. Arcata's project should be looked at as part of a regional answer, an answer that demonstrates innovative thinking on the part of government . . . The expense of the huge single system should never be justified because of administrative convenience or because the larger the project, the more quickly federal and state moneys will be available to build it. The ease of governmental life should never be a reason to make citizens pay more than is necessary."

There was a brief, stunned silence after Rose finished speaking. Then John Stokes, the attorney for HBWA, began to interrogate Rose in the manner of an outraged headmaster. He accused Rose of malingering with known enemies of the HBWA project, such as Jackie Kasun. He uncovered the fact that Rose's formal training had been in psychology rather than wastewater engineering. "You have no formal qualifications to make any statement on Arcata's project, is that correct?" Stokes snapped.

"I've been studying the wastewater process extensively for over a year now," Rose answered. "As a person, I can learn and read." Rose, a new state employee, had dissected the psychology behind the push for regional sewage systems, and defied the word of the experts—a subversive act that outraged not only Stokes but many long-time state bureaucrats. Later in Humboldt County's wastewater wars, when he traveled to speak on behalf of Arcata's marsh alternative, he'd be dogged by state operatives in suits and dark glasses, who implied dire consequences if he spoke.

At the close of the hearing, the regional board's attorney restated the Bays and Estuaries policy as the letter of the law. Andrea Tuttle, the young board member from Arcata, spoke in a tight voice. "I feel like we're under a steamroller right now. I'm going to have to pay for this plant, and I also have to live with these people as friends, and I hope I can maintain their friendship. I find the regional plant very hard to stomach. I dislike the cost, the land use planning ramifications, I dislike piping raw sewage under the bay. The problem is that marsh treatment is an untried technology. As a friend, I would vote for Arcata, but as a responsible member of this board I cannot do it."

When the formal vote was called, every board member expressed sympathy for Arcata, and every board member voted the city's proposal down. By the time her name was called to vote, Tuttle had left the room in tears.

Hauser was unsurprised by the board's decision, and ready to fight on. "We fell in love with the idea of building a marsh," he explains. "If they'd compromised early on, the regional plant probably would have been built." The vision of a revived wetland changed everything for the Arcata contingent and made them committed to defeating HBWA any way they could. "It came to the point," reflects Hauser, "where we were just going to kill that sucker."

NOTES

1. Allen, G.H. (December 1, 1970). "The constructive use of sewage, with particular reference to fish culture." *FAO Technical conference on marine pollution and its effects on living resources and fishing*. FIR: MP/70/R-13.
2. Scott, E., Leslie Brunetta (1989). "Wastewater wars." *Kennedy School of Goverment Case Program*, C16-89-854.0.
3. Kasun, J. "The Humboldt Bay Wastewater Authority East Bay Interceptor: an analysis." *Humboldt State Library.* Humboldt Room collection.
4. Anonymous (October 14, 1976). "Bonds vote bid lost." *Times Standard.*
5. Anonymous (November 23, 1976). "Letter may peril HBWA grant hopes." *Eureka Times Standard.*
6. Bretnall, P. (1984). "Wastewater conflict on Humboldt Bay." *Humboldt Journal of Social Relations* **11**(2): 128–284.
7. Scott, E., Leslie Brunetta (1989). "Wastewater wars." *Kennedy School of Goverment Case Program*, **C16-89-854.0**.
8. Anonymous (May 27, 1977). "Arcata will get chance to be heard by WQCB." *Times Standard.*
9. Ibid.
10. Van der Ryn, S. (1978). "The Toilet Papers." *Santa Barbara: Capra Press.*

7

The United States of Vanished Wetlands

Before he became a revolutionary general and the nation's first president, George Washington was a destroyer of wetlands. In 1763, he surveyed the edges of a million-acre expanse of wet forest that lay along the Virginia–North Carolina state line. He described the Great Dismal Swamp as a "glorious paradise" full of wildfowl and game.[1] Still, he seemed to have no qualms about dismantling Eden. In 1764 he applied with five partners for a charter to create a business called "Adventurers for draining the great Dismal Swamp."[2] Their goal was to chop down and sell the timber from majestic cypress and cedar trees, then to plow the land for crops.

The brutal work of digging drainage ditches and canals was done by slaves. By the time of the Revolutionary War, the Adventurers Company was producing 8 million shingles a year for sale—valuable slivers of wood cut from the swamp's enormous bald cypress trees.

There was profit in undoing wetlands. Draining a wetland also seemed to make a place healthier. People who colonized swampy land were plagued by a dreadful illness, one that often killed, and left survivors with recurring bouts of a bone-rattling fever. Malaria—the name itself means "bad air"—was believed to be triggered by poisonous vapors rising from still waters.

The drainage and destruction of wetlands was an unwritten founding principle of the US. The pattern began with some of the earliest European settlers. Well before the colonies won their independence, the loss of wetlands had led to pollution that changed the ecology of rivers and bays. Over the centuries, wetlands loss and water pollution have accelerated in tandem, driven by the need for farmland, the urge for profit, and the fear of disease.

The history of these interwoven changes on land and underwater begins in the Chesapeake Bay, the site of the first permanent British colony in America.

In the summer of 1608, Captain John Smith and the colonists of Jamestown were starving. As they sailed into the Chesapeake Bay for the first time, their guts ached and their bones stood out in sharp relief. In the bay they found salvation: a dazzling array of edible creatures. At ebb tide, rich beds of mussels and oysters were uncovered. Blue crab scuttled in the shallows, armed with turquoise claws. Atlantic sturgeon surged upstream to spawn; an adult could weigh more than eight hundred pounds. The waters were thick with shad and striped bass. A member of the crew wrote that none of them had ever seen such aquatic bounty before.

Captain Smith would not stay long. Injured in a gunpowder explosion, he returned to England for treatment in 1609 and never saw Virginia again. He advertised the wonders of the Chesapeake region, however, helping to recruit new colonists. “Heaven and Earth never agreed better to frame a place for man’s habitation,” he wrote, “were it fully manured and inhabited by industrious people.”

Smith and his men would not recognize today’s Chesapeake. On a typical summer’s day, 60 percent of the bay’s waters are a dead zone, devoid of oxygen, where no fish or crabs can survive. Sturgeon have become a rarity, shad fisheries are closed, striped bass are in decline, oysters have all but vanished. The estuary suffers from an overload of nitrogen, phosphorus, and sediment, generated by the 17 million industrious people who now live in the Chesapeake watershed.

The bay’s watery world has been transformed by the axes and plows, and later the bulldozers and cement mixers, of people on land. Paleoecologist Grace Brush has used plant remains preserved in the muddy bottom to read the story of drastic change in the bay’s underwater communities. These relics have given her an intimate knowledge of the Chesapeake the first colonists encountered. She has tracked the beginnings of the ecosystem’s decline to the early years of European settlement.

When the first colonists arrived, the land surrounding the bay was covered in forest and rich in wildlife. Beaver were abundant. The big rodents cut down trees using their powerful buck teeth, and built dams that formed marshes. Beavers shape the nature of a watershed: Their dams hold not just water, but rich stores of nutrients that would otherwise run off the land. Boggy beaver country made an ideal habitat for the soil microbes that process nitrogen, changing it into forms used by plants and animals, then pulling it out of the realm of living things and back into the air as N_2 gas.

Beaver furs were a prized commodity in Europe. British colonists feared the swamps where beaver lived, so they traded with the Indians for pelts. Beginning in the 1620s, hundreds of thousands of beaver were trapped in Connecticut, Massachusetts, and New York, and on the streams feeding the Chesapeake. In 1638, demand increased when King Charles II made the use of beaver pelts mandatory in hat making.[3] By the mid-1700s, beaver were gone from the Chesapeake watershed.

The beavers’ disappearance left a signature in bay sediments: After 1750, levels of nitrogen dropped, possibly due to a sudden absence of beaver poop.[4] Later, after beaver dams slowly rotted away, ponds drained, and areas of wet soil that hosted denitrifying bacteria dwindled. Increasing loads of nitrogen began to run into the bay.

The pulse of added nutrients is recorded in the remains of centuries-old algae. Diatoms are single-celled green algae that form skeletons of silica—each species leaves behind a skeleton with a distinctive shape. Under the microscope, these may resemble a five-pointed star, a pasta shell, a wagon wheel, or a truck tire. In the 1700s, concentrations of nitrogen and sediment climbed, and the diversity of diatom populations began to dwindle. The *Nereis* worm, a creature that fed on the bay bottom and was a favorite prey of shad and other native fish, had been abundant. After the beaver were hunted out, *Nereis* worms began to die off.

Meanwhile, settlers were chopping down trees and plowing the land to plant wheat and tobacco. The crops exhausted nutrients in the soil, so farmers imported guano from South America as fertilizer. The load of nutrients running off the land increased.

Before European settlement, the Chesapeake's bounty had relied on bottom-dwelling organisms. These included aquatic grasses and diatoms that grew on the sediments or on the leaves of larger plants. As they transformed the sun's energy into green growth, the underwater plants pumped oxygen into the depths.

By the late 1800s, forests and wetlands were gone from the watershed, translated into tidy farm fields or buried under city streets. The release of nutrients from land transformed life in the estuary. The waters were clouded with blooms of floating algae, which blocked light. Aquatic grasses faded, and with them went the benthic diatoms and *Nereis* worms. Oysters vanished, victims of pollution and overfishing. The bay's bottom turned barren, depleted of oxygen. The entire aquatic ecosystem was turned inside out. Once benthic and diverse, it became planktonic, able to support only a few kinds of plants and animals adapted to live in the upper layer of over-fertilized waters.

Onshore, settlers' traditional fear and loathing of wetlands was reinforced. Many suffered from a disease they knew in awful detail. The illness was called "marsh fever" or "ague," and everyone knew the symptoms: a fever that returned every third or fourth day, bringing on hot sweats followed by cold shakes; a painful swelling of the spleen; weakness; and a yellow complexion.

Malaria was carried to America in the veins of European colonists, and later in those of the slaves they brought from Africa. Many of the early Chesapeake colonists came from the marshy parts of Britain: Kent, Sussex, and the shores of the Thames, where malaria had been endemic since the fifteenth century. People there believed the illness was caused by bad air emanating from wetlands. The impoverished parish of Romney Marsh had one of the highest mortality rates in eighteenth-century England. "The large quantity of stagnating waters engenders such noxious and pestilential vapours as spread sickness and frequent death on the inhabitants," wrote a visitor.[5]

A high proportion of children born in seventeenth- and early eighteenth-century Virginia died between one and four years of age. Babies receive short-term immunity to malaria from their mother's milk, but this ends after the first year, and many young children can't fight the disease off on their own.[6] The colony's death records echo modern mortality statistics from sub-Saharan Africa, where malaria remains endemic and the great majority of malaria deaths strike children under the age of five.

Many of the settlers must have been infected, or malaria would not have taken hold in the New World. The disease involves intricately timed interactions between protozoan parasites, short-lived mosquito hosts, and the humans adult female mosquitoes feed on. Only mosquitoes in the genus *Anopheles* carry malaria, and the species that transmit the human disease are adapted to live among people, resting on the walls of houses and laying their eggs in a puddle as small as a hoof print. Malaria did not exist in pre-settlement America, but an eligible mosquito species, *Anopheles quadrimaculatus*, was waiting on the Chesapeake shore to receive the parasite.

Malaria persisted. It would travel west with new waves of immigrants, cause great suffering and early death, and add to settlers' intense dislike of wetlands.

A few swathes of wet wilderness were so vast, mucky, and difficult to cross that they foiled the efforts of pioneers to tame them, at least for a time. Among these remarkable places was the Great Black Swamp, an expanse of wet forest and marsh that stretched across a million acres in northwest Ohio and Indiana. The wilderness was home to wolves, deer, elk, black bear, mountain lion, beaver, and river otter. Now it is gone, the land where great beech, elm, and sycamores grew parceled into farm fields.

In 1794 US soldiers under the command of General Anthony Wayne, a Revolutionary War hero called out of retirement by President Washington, marched along the upper Maumee River past a string of Indian villages. Wayne marveled at the "very extensive and highly cultivated fields" that lay along the river's edge. He'd never seen such immense fields of corn, he said, in any part of America, from Canada to Florida. He had his troops set the Indians' homes and impressive crop on fire, then they marched on downriver.

Wayne's troops moved through parts of the Black Swamp and camped in the wetland on August 19. The next day, at the Battle of Fallen Timbers, just south of present-day Toledo, they defeated the Western Confederacy, an alliance of tribes that had joined forces to defend their lands. The soldiers returned upriver to Fort Defiance, where many of the men fell sick with malaria.

Malaria was a deadly threat to Indians, who had no ancestral immunity to the parasite. While bivouacked in the Black Swamp, US Army Major B.F. Stickney daydreamed about using the place as a biological weapon against the Indians. Force the natives to live in the swamp, bring them plenty of good food, keep them there for six weeks, and those who didn't die would be too weak to fight. "All the lives of the troops would be saved, and at least three-fourths of the cash," he wrote. Stickney admitted there might be some "question of morality" in this tactic.[7] Though the army never forced native people into the swamp, many Indians who survived clashes with the army would die of malaria.

After Wayne's victory, the tribes were forced to cede their lands, but settlers were slow in coming. The Swamp was nearly impossible to move through, and the region had a reputation for unhealthiness. At the outbreak of the War of 1812, the only white settlements in northwest Ohio were at the rapids near the mouths of the Maumee River, on the swamp's western edge, and the Sandusky to the east. About four hundred settlers and three thousand Indians lived in the region. When Detroit fell to the British, the settlers fled, and troops were sent under the command of General William Hull to retake the city.

To get to Detroit, Hull's men had to slog through the dark heart of the Black Swamp. It was spring, when the swamp was at its wettest. Soldiers staggered along, thigh-deep in mud, with frequent stops to pry out horses and wagons stuck in the mire.[8] At night they pitched tents, but the mud collected ankle-deep inside. Journals kept by the men under Hull's command record miserable, wet nights plagued by mosquitoes. Soon after, they suffered the fever and weakness of malaria.[9]

The swamp was "astonishingly fruitful in the production of marsh miasmata," wrote a regimental surgeon.[10] Encamped in the cold waters of the swamp, many of the sick died. One soldier, Elias Darnell, wrote that staying in the swamp killed more of his comrades than an intense battle.

Yet the Black Swamp and Lake Erie's southwestern shore were also bountiful. The Maumee ran clear and was thick with fish: bullhead, bass, sturgeon. On an April morning in 1813, two hungry soldiers stationed at Fort Meigs, near present-day Toledo, walked down to the Maumee River. The clear waters swarmed with perch, muskellunge, sturgeon, and catfish. Plunging spears into the water at random, they caught sixty-seven fish in thirty minutes, often killing two or three with a single stroke.

Every river mouth west of the Sandusky held dense beds of wild rice, where waterfowl settled to feed, then rose in flocks that darkened the sky. The stalks could stand higher than a man's head: To feed, ducks grabbed the stems with their feet and tugged the seedheads down to the water. Waterfowl were so abundant, so fearless and loud, that their constant quacks and honks kept a nervous young army recruit lost in the marsh awake all night.

After the war ended in 1815, white pioneers trickled back to northwest Ohio, settling on the fringes of the Black Swamp, in the same places where Indian villages had stood. Many more emigrants wanted to pass through on their way to Michigan, Indiana, or Illinois, but the swamp remained a formidable barrier.

The first road across the swamp was funded by Congress and completed in 1827 after years of labor (Fig. 7.1). The Black Swamp Road was a thirty-one-mile ribbon of soggy earth, ditched on both sides, cut through forest so dense that travelers never saw the sun. It earned a reputation as the worst route on the continent.[11] The more wagons that sloshed over it, the more impassable it got. Nevertheless, it was used by growing numbers of settlers heading west. In the 1830s, locals earned steady money consoling emigrants and hauling stalled wagon teams out of the worst mud holes. Thirty-one taverns stood along the road between Fremont, on the Sandusky, and Perrysburg, on the Maumee, an average of a tavern for every mile of road.

By 1850, Congress was addressing the widespread urge to do away with all the nation's great wetlands. In a series of new laws collectively known as the Swamp Land Acts, states were granted rights to sell off wetlands within their borders. The idea was that private owners would drain the land, transforming swamps and marshes into productive farms.

In the Black Swamp, the choice farm lands were along streambanks, which formed strips of dry land that could be cleared of trees and farmed without artificial drainage. Inland from the streams, the soil was saturated with water and crops withered. A few German and English farmers settled in the swamp, cleared their plots, and dug ditches to channel excess water off their fields. The advantages of this technique soon became obvious. The land was so flat, however, that workable drainage ditches had to run a long distance, and could not be dug without crossing a neighbor's property.

Figure 7.1 A crew clearing trees to build a railroad extension across the Great Black Swamp. Photo from Wikimedia.

The local historian Homer Everett, who witnessed the settlement era, wrote that a Yankee would refuse to let a German neighbor dig a ditch through his land, though the drainage would have helped them both. Prejudice and petty disagreements stood in the way of drainage, and therefore of successful farming. "If Mr. Johnson owned a piece of wet land near Mr. Jones, and wanted to get the water off by draining through Jones' land," wrote Everett, "he could not obtain it because, perhaps, Johnson, ten years before, threw a club at Jones' yellow dog to drive him out of the road and keep himself from being bitten."[12]

In 1859 the Ohio General Assembly passed a law authorizing county commissioners to construct drainage ditches. Farmers benefiting from ditch construction shared the cost. The other Midwestern states also enacted laws authorizing drainage districts, enabling the construction of ditch networks that drained great swathes of land—a mission that required investment and coordination, and could not have been accomplished by individual landowners.[13] Through the work of drainage districts, the Corn Belt states of Ohio, Indiana, Illinois, and Iowa would lose more than 95 percent of their native wetlands.

At first, work in the Black Swamp focused on clearing, deepening, and widening natural channels, which were so cluttered with log jams and beaver dams that water barely moved through.[14] Construction of new drainage ditches came later. The most dramatic example is the Jackson Cutoff in Wood County, which is seven miles long and more than twenty feet deep in places. The cutoff was dug in 1879 at a cost of $110,000. First timber crews cleared the trees, saving enough lumber for bridge construction and burning the rest. Work horses pulled plows through the ground, followed by a second crew that dug the channel using horse-drawn scrapers. Men with shovels did the rest of the digging, removing the heavy clay that lay beneath the rich topsoil. The Jackson Cutoff drains thirty thousand acres in Wood County, emptying into Beaver Creek near its confluence with the Maumee—and it is just one ditch among hundreds.

Yet even this proliferation of ditches was not enough to keep farm fields drained and productive. For the first few years after a plot of forest had been cleared, decaying tree roots underground funneled water out of the topsoil. Once the roots had rotted away, the channels they'd formed collapsed. The soil became waterlogged and depleted of oxygen, and crops failed.

To survive, farms in the region needed underdrainage: a series of channels laid beneath a field that would carry excess water into the nearest ditch. The best way of doing this was with drainage tiles, short pieces of ceramic pipe. Laid end to end in a furrow, then covered over with soil, these tiles could function for many years. The technique had been brought to the US in 1821 by John Johnston, a Scottish immigrant farming in Geneva, New York.[15] Though his neighbors mocked him at first, Johnston's ability to turn unusable, water-soaked land into productive fields soon changed their minds. Influenced by Johnston, a Geneva potter named B.F. Whartenby patented the first tile-making machine in America.

Wood was cheap and abundant in the Black Swamp—there were more than four hundred sawmills operating in the region by the 1870s. The nearest tile factory was far off in central Ohio, so farmers used wooden underdrains, made by nailing planks together in a V-shape. Underdrainage raised crop yields, and the demand for affordable clay tile grew. Some enterprising soul tested the abundant clay that lay a foot or two beneath the surface of the Black Swamp and found that it made excellent tiles. By 1879 there were eleven tile factories in Putnam County alone, producing eighty thousand feet of tile per year.[16] In 1880, more than fifty tile factories operated in northwest Ohio.

The swamp, once a forbidding and near-impassable wilderness, was dismembered and used to feed an accelerating cycle of human industry. The great wetland trees—ash, elm, sycamore—were felled and used to build houses, make furniture, and fuel the railroads that sprouted up across Ohio. In the 1860s Ohio's railways consumed 1 million cords of wood each year as fuel, and an unknown quantity for ties. The process of railroad construction involved building drainage ditches, and showed settlers that the mucky soil could be reclaimed for farming. The discovery of underdrainage created a growing demand for tile. All this drove an orgy of forest-clearing and land-draining, which in the course of five decades (from 1870 to 1920) completely erased the Black Swamp, leaving an orderly landscape of farm fields in its place. A wilderness went up in the smoke from railroad engines, and flowed in drainage ditches down to the Maumee, which began to run murky instead of clear.

The Black Swamp's undoing was speeded by James B. Hill, a native of Fremont, at the eastern edge of the swamp. Born in 1856, when most of the swamp still stood, Hill experienced firsthand the back-breaking labor of digging ditches to place tile. In 1893, he built the first successful steam-driven tractor ditcher while working in a machine shop in Bowling Green, a town set in the heart of the swamp. Hill's invention, known as the Buckeye Traction Ditcher, could set tile faster and more accurately than a crew of experienced men. In 1905, a Buckeye Ditcher raced fifty hand ditchers and laid four hundred feet of perfect trench to the hand crew's three hundred feet.[17] The Buckeye Ditcher would help to place thousands of miles of tile in the Black Swamp, and would be used to drain the Florida Everglades and large stretches of Louisiana wetland.

In 1858, as early settlers struggled in the mud of the Black Swamp, George Waring engineered a network of drainage tiles on a stretch of marshy ground in the heart of Manhattan. Following his plan, workers hand-buried more than sixty miles of clay pipe, draining the land so that Central Park could be created.[18] Waring did such an expert job that more than a century later, some of the original clay tiles were still carrying water and keeping the ground dry.

Waring's work on Central Park was the largest drainage project of its time, and made his reputation. He started out as a farmer and ended up as one of the most respected sanitation experts in the US. He was a drainage fanatic and a staunch believer in the notion that wetlands poisoned human health. "Land which requires draining hangs out a sign of its condition," he wrote. "Sometimes it is the broad banner of standing water, or dark wet streaks in plowed land, when all should be dry and of even color . . . sometimes the quarantine flag of rank growth and dank miasmatic fogs."[19] Waring personified the era's attitudes toward wetlands, farming, and public health (Fig. 7.2). He lived by the notion of killer swamp miasmas—and died by it.

In 1861, when his work on Central Park was almost completed, Waring joined the Union Army as a cavalry officer, and rose to the rank of colonel. After the war he managed a Rhode Island farm and wrote, everything from books on scientific agriculture to popular horse stories and accounts of European travel. He tried, and failed, to market a design for an earth closet, a device that might have limited the gush of water pollution that came with the era of the flush toilet. He wrote a popular and influential book, *Draining for Health and Draining for Profit*, which summed up his view of wetlands as cauldrons of disease, and gave detailed instructions on how to make them disappear.

By the 1870s, Waring had become a sanitary crusader, warning Americans of the danger of sewer gas, the stink that arose from badly plumbed water closets and backed-up sewers.[20] He argued that marsh miasmas and sewer gas were the source of every contagious disease. Waring would carry the banner for miasma even after Pasteur, Lister, and Koch had demonstrated the existence of pathogenic microbes. Many physicians of the time dismissed the newfangled germ theory of disease and agreed with Waring; his 1878 essay on the miasmatic origin of typhoid fever was awarded a prize by the Rhode Island Medical Society.

That year a devastating epidemic of yellow fever struck New Orleans. Half the city's population fled, catching any train they could. These refugees brought the infection to other cities: Mobile, Chattanooga, Memphis.

Yellow fever is a particularly terrifying disease. The first signs of infection are aches and fever, followed by jaundice. The liver breaks down, destroying proteins needed for blood clotting. Victims bleed from their eyes, nose, and mouth. In the final stages, they bleed into their stomachs and vomit up partially digested blood. Death came in a frantic agony, and its source remained mysterious.

Bacteria had been identified as the cause of cholera, but yellow fever is caused by a virus and transmitted by mosquitoes. Viruses were then unknown entities, invisible under nineteenth-century microscopes. The concept of insects as disease vectors was unimaginable.

The 1878 epidemic spread to eleven states, infecting 120,000 people. It ended with the arrival of the first frost in autumn. By then twenty thousand people had died, five thousand of them in Memphis.[21]

Acting as a special commissioner by presidential appointment, Waring went to Memphis to sanitize the city as a protection against future outbreaks. The city had no sewer system. Its streets were full of filth, and Waring saw that filth as the source of the epidemic. He designed and supervised the building of a sewer system that kept human waste and storm runoff separate, an innovation that would be adopted nationwide. Like many of his ideas, the concept of the separated sewer system was adopted from Edwin Chadwick's work in England. Memphis remained free of yellow fever, and Waring's efforts seemed a great success.

Demand for his services grew. Waring acted as a consulting sanitarian for the wealthy landowners of Newport, Rhode Island, and later supervised the cleanup of New York's befouled streets, enlisting local children as his foot soldiers. He became a public health hero, "the apostle of cleanliness, the scourge of dirt."[22]

When the US declared war on Spain in 1898, forty thousand American troops landed on the Spanish colony of Cuba. Yellow fever was endemic on the island. In only three months, the US Army defeated Spain and occupied Cuba. Fewer than four hundred US soldiers died in combat, but more than two thousand were infected with yellow fever.[23]

Waring was asked to chair a commission of experts charged with cleaning up Havana and stamping out contagious disease. He traveled to Havana in October 1898, where he found a city without sewers, where garbage and feces were strewn in the streets. There was a good water supply, but Havana was surrounded by marshes, which he saw as generators of miasmatic illness.

Waring left Havana after three weeks and on his way back to the US wrote up his plan for the city's rescue. He would ban privies, replacing them with flush toilets. He would build a sewer system, pave the streets, drain the marshes. He believed his plan would put an end to yellow fever and other "miasmatic" diseases in Havana, at an estimated cost of $10 million.[24]

Waring fell sick the day after he returned to New York. Four days later, he died of yellow fever. The miasma theory would die with him.

William Gorgas, the US Army doctor in charge of sanitation in Havana, shared Waring's belief that yellow fever was caused by filth, and followed much of Waring's advice. Havana was transformed into one of the cleanest cities in the Americas. Despite all this, the spring of 1899 saw a fresh outbreak of yellow fever: Cleanliness, it seemed, was not enough.

The man who solved the puzzle was a young bacteriologist, Jesse Lazear, who was hired to head the Army's disease lab in Havana. Lazear had been studying malaria and knew that breakthrough research in India had just shown that the disease was transmitted by mosquitoes. Ignoring the skepticism of his colleagues, he consulted with a Cuban doctor, Carlos Finlay, who had painstakingly matched the pattern of yellow fever infection in Cuba to the presence of a single mosquito species, *Aedes aegypti*.

Figure 7.2 George Waring in 1883. Photo from Wikimedia.

Lazear raised *A. aegypti* mosquitoes in his lab, then had adults feed on infected patients in the Army's yellow fever ward. Healthy volunteers agreed to be bitten by the fever-exposed mosquitoes. The experiment took time and repeated trials, because the yellow fever virus must incubate in a mosquito for several days before it becomes contagious. Eventually, the lab-raised mosquitoes did cause yellow fever in volunteers. Several of them died, including Lazear himself.

To prove beyond doubt that yellow fever was transmitted only by mosquitoes, Walter Reed, head of the Army's Yellow Fever Commission, had volunteers sleep in the pajamas and on the sheets used by yellow fever victims, soaked in their vomit and feces. None became ill.

This dramatic evidence convinced Gorgas: He abandoned the miasma theory and made war on mosquitoes. His tactics were efficient but practical only under a state of martial law. He sent sanitation squads into every corner of Havana, coating every puddle of standing water with oil (a thin layer of oil kills mosquito larvae). He prowled the streets, personally checking that all residents oiled or screened every bit of water on their property. *A. aegypti* breeds in the small bodies of water that surround humanity: the water-filled hoof print of a cow, a bucket left standing, a household cistern.

Gorgas would later use similar techniques during construction of the Panama Canal in the early 1900s. Decades earlier, when French colonialists had attempted to build a canal across the isthmus, twenty-two thousand workers died of mosquito-borne disease, halting construction. If workers dropped at the same rate, America's canal would never be completed.

Gorgas fumigated every building in Panama City and oiled the streets. He had his crews attack all potential mosquito breeding grounds in a long, narrow strip of jungle surrounding the construction zone. They dosed any patch of standing water with a mixture of carbolic acid, resin, and caustic soda. Workers slept under bed nets, and servants patrolled the railcars where workers slept, armed with fly swatters and bottles of chloroform, killing mosquitoes one by one.

These tactics focused on the *A. aegypti* mosquito and, in 1906, succeeded in extirpating yellow fever in Panama.[25] Malaria, however, remained an intense problem, affecting the majority of canal workers. Little was known about the ecology of malarial mosquitoes. That changed when a US Department of Agriculture (USDA) entomologist, August Busck, visited Panama. He collected mosquito larvae from throughout the region and reared them to adulthood. Busck identified more than ninety species, thirty of them unknown to science. His new insights into mosquito ecology made it clear that construction work on the canal was creating habitats for the species most likely to act as malaria vectors: Larvae grew in the puddles between railroad ties, and in ponds formed where excavation spoils were dumped.

The main malaria vector in Panama proved to be *Anopheles albimanus*. Highly susceptible to malaria parasites, *A. albimanus* seeks out humans as a source of blood and breeds in the small puddles that surround human habitation. Joseph LePrince, who worked as Chief Sanitary Inspector on the canal project, came to know the species on intimate terms. He invented a way of tracking mosquito movements by coloring adults with blue aniline dye, a tactic that showed the mosquitoes traveled no more than a mile from their hatching sites.[26] This discovery let public health workers focus on breeding habitats within a mosquito-flight of the work crews building the canal.

LePrince brought the mosquito-fighting strategies developed in Panama back to the US. During World War I he was in charge of mosquito control at military bases in the South, where malaria remained a serious problem, and succeeded in dramatically lowering infection rates. He identified *Anopheles quadrimaculatus*, later known as the quad mosquito, as the vector of malaria in North America. After the war he continued to battle mosquitoes as an officer of the US Public Health Service (USPHS), using strategies that included draining marshes, stocking ponds with minnows that devour mosquito larvae, supplying low-cost screens for rural households, and spraying the shallow waters of lakes with Paris green, a highly toxic compound of copper and arsenic.[27]

The new insight into the link between mosquitoes and disease triggered a flurry of wetland drainage. In the 1910s, the New York City Health Department hired hundreds of workers each summer to dig drainage ditches and build tidal gates in coastal marshes. New Jersey followed, requiring local governments to create their own "county mosquito extermination commissions."[28] In the South, the USPHS took charge of drainage campaigns against mosquitoes.

Ditching of salt marshes was meant to increase tidal flow and prevent the pooling of stagnant water that mosquitoes need for breeding. The work was done at random, however, without testing for the presence of quad mosquitoes. Poorly done drainage efforts often created new puddles of standing water where mosquitoes could breed. There's no clear evidence that the mass drainage efforts of the early twentieth century helped to control malaria. A more certain result of ditching was the loss of native salt marsh plants, the lowering of the water table, and the destruction of habitat for birds and fish.[29]

Epidemic malaria had slowed the settlement of southern Illinois and Missouri and plagued settlers throughout the Midwest, but it was gone from the region by 1890. People assumed this was a happy effect of wetland drainage.Yet the disappearance of malaria in the Midwest cannot be traced to a single cause. Drainage of wetlands for agriculture was not the most significant factor—it could not have been, because malaria began to dwindle years before the establishment of organized drainage districts.[30]

In American history, malaria was very much a disease of pioneers. As masses of people moved west, they carried their malaria parasites with them, just as early Chesapeake colonists brought their parasites from England. The log cabins of early settlers were crowded with a dangerous combination of immune malaria carriers, non-immune people vulnerable to infection, and hungry mosquitoes. Quad mosquitoes found plenty of shelter in the dark crevices of log cabin walls. As human populations stabilized in the late nineteenth century, people stopped moving and started living in better housing, and infection rates went down.

Other changes helped too. Railroad construction carried people inland, allowing them to settle away from mosquito-infested bottomlands. The rise of dairy cattle in Iowa, Wisconsin, and Minnesota gave quad mosquitoes an alternate source of blood meals, decreasing human infection rates.

In the colder parts of the country, where temperatures were less favorable to malaria parasites and mosquitoes could be active only a few months of the year, the disease faded as settlements became more established. In New England and New York, malaria was first reported in the 1670s and was gone by 1750. The disease remained endemic in the American South, where warmer temperatures favored the persistence of malaria parasites and mosquito vectors remained active most of the year. It would migrate North again during and after the Civil War, when thousands of soldiers infected in boggy Southern battlefields returned, carrying malaria home with them.

In the summer of 1905, a young ornithologist and wildlife photographer named William Finley ventured into the wet wilderness of the Klamath Basin on the Oregon–California border. In Lower Klamath Lake, he and his partner Herman Bohlman found thousands of water birds nesting on floating islands of tule, or hard-stem bulrush (Fig. 7.3). The ten-foot-tall, tubular leaves of tule had formed flexible mats over centuries of growth, the green shoots of each spring weaving up through the dead stalks of previous years.[31] Finley and Bohlman lived among the birds on these precarious tule mats for two weeks. At times the surface of the floating islands gave way without warning, dumping the men into the lake. They laid their sleeping bags out on piles of tule straw three feet deep, slept through the night, and awoke floating just at the water's surface.[32]

Surging beneath them was an astonishing wealth of lake trout and suckers. At one fishing spot where the rocky bottom forced migrating suckers to swim to the surface, Klamath and Modoc Indians had caught more than fifty tons of suckers each year until 1900. Tule Lake, just east of Lower Klamath, hosted one of the largest breeding colonies of osprey (fish-eating hawks) in North America.[33]

Finley and Bohlman captured a series of intimate portraits of breeding pelicans, avocets, western grebes, terns, and ducks. Finley published a written and photographic portrait of bird life in the tule jungles of the Klamath Basin. He brought his photos to the White House, inspiring President Theodore Roosevelt to issue an executive order in 1908 setting aside all of Lower Klamath Lake—eighty-one thousand acres of lake, islands, and marsh—as a wildlife refuge. It was the first such refuge in the West created with the goal of protecting waterfowl.

The birds were in dire need of protection. Plume hunters sought out adult egrets, terns, gulls, grebes, herons, and pelicans on their nesting grounds and killed them by the thousands. In a single summer, hunters shipped thirty thousand grebe skins to San Francisco from Klamath Lake. Hunters also shot thousands of ducks: In 1903, more than 120 tons of ducks were shipped to San Francisco meat markets from the Klamath Basin. Plume hunters had already devastated the once-abundant bird populations of the Florida Everglades.

The Oregon Audubon Society lobbied the state legislature to pass a law against killing "inedible" birds, like herons and pelicans. The National Audubon Society paid the wages of two state game wardens to enforce the new law on Lower

Figure 7.3 William Finley sitting at an umbrella blind, taking notes, during his 1905 Klamath expedition. Several gull chicks are perched around his legs. Photo Oregon Historical Society, Org Lot 369, Finley A1600.

Klamath and Tule lakes. Yet the federal government would prove a deadlier enemy of wildlife than the most unscrupulous hunters: At great expense, federal engineers would drain Lower Klamath Lake dry.

The story of the Klamath Refuge illustrates the long conflict between conservationists and the US Reclamation Service, which drained wetlands and diverted rivers throughout the arid West. Created with the support of President Roosevelt under the Reclamation Act of 1902, the Service's mission was to turn unproductive land into small, irrigated family farms. Roosevelt had declared the Lower Klamath Lake Refuge on a rich wetland already slated for destruction.

The Reclamation Service proposed to build two dams, which would cut off the flow of water to Lower Klamath and neighboring Tule Lake, exposing their beds for farming. During the summer dry season, irrigation water would be released from the dams. The arrival of Reclamation Service engineers in 1904 set off a land rush, in which speculators snatched up large plots of still-wet lakebed and marsh at low prices, intending to resell at a profit after the reclamation project was complete. By 1908, when Roosevelt declared Lower Klamath Lake a wildlife refuge, one-third of its acreage was already in private ownership. Over time, more and more of Lower Klamath and Tule Lake were transferred into the hands of homesteaders and speculators.

Under the law, federal reclamation was meant to provide affordable land for family farmers. Frederick Newell, head of the Reclamation Service, insisted that the government would not invest a single dollar in the Klamath Basin until lands there were sold off in small units of 160 acres or less. While land speculators complained bitterly that this deprived them of their private property rights and refused to sell, Reclamation Service engineers went ahead with design and construction of the dams. The way events played out in the Klamath Basin was typical. Despite the letter of the Reclamation Act, on the ground federal engineers acted to enrich speculators, not to provide affordable new farms for impoverished families. The Act's acreage-limitation rule went routinely unenforced; a 1916 review board concluded that the government's management "closely verges on fraud."[34]

Reclamation Service engineers saw it as their mission to re-plumb the West in the interests of agriculture. Their indifference to the details of the law that had created their agency was fueled in part by self-interest: Under contracts with landowners' associations, reclamation projects brought in large amounts of money to the Reclamation Fund that paid their salaries. The Reclamation Service would grow to become the largest bureaucracy in the history of irrigation and worked at cross-purposes with the underfunded US Biological Survey. The Biological Survey had no money to pay game wardens, so the Audubon Society and the State of Oregon came up with the cash. In a pattern that would be repeated in other federal refuges, the first Klamath Refuge wardens were defied by the local district attorney, who announced that the feds had no jurisdiction and he'd continue to hunt as much as he wanted.

Construction on the Klamath Project began before any data had been collected on conditions in the basin. The lakes and marshes were watered by melting snow that ran off the surrounding mountains, giving the region a deceptively lush appearance. Eventually federal officials noticed that the basin made a poor place to farm—chilly, with little rainfall and a short growing season. A 1909 test by the USDA found that crops would not grow in a drained section of Lower Klamath marsh because the soil was full of alkaline salt, but by then it was too late to stop the project's momentum.

On November 30, 1917, a group of private landowners who had organized as the Klamath Drainage District signed a contract with the Reclamation Service. That day the headgates of the Klamath Dam closed, cutting off the flow of water to Lower Klamath Lake. By 1919, the lake had been reduced to a wide, shallow puddle. It still attracted birds, but they died in large numbers of alkali poisoning. A few years later, all that remained of the bountiful wildlife habitat Finley had once explored was a 365-acre sump at the southern end of the lake bed.

Finley, who by the time the headgates closed had been appointed State Biologist for Oregon, loudly protested the destruction of the Lower Klamath refuge. Two years later, he was fired without notice by the Oregon Fish and Game Commission. Several of the commissioners were boosters of the Klamath reclamation project and were fed up with Finley's passionate campaign to reflood the refuge.

Soon after, cattle ranchers on the south end of the old lake bed complained that the water table had sunk, drying out their pastures. The vanished marshes had left behind a thick layer of peat, which dried out in the sun and caught fire. In the mid-1920s, Finley described the desolate landscape:

> Today, Lower Klamath Lake is but a memory. It is a great desert waste of dry peat and alkali. Over large stretches fire has burned the peat to a depth of from one to three feet, leaving a layer of white loose ashes into which one sinks above his knees. One of the most unique features in North America is gone. It is a crime against our children.[35]

Among those who joined Finley in demanding Lower Klamath Lake's revival were duck hunters, naturalists, and residents of nearby Merrill, Oregon, who suffered from repeated dust and ash storms blowing off the dry lakebed. In the fall of 1922, schools in Merrill and Klamath Falls were closed when a thick cloud of ash enveloped both towns. It was a small-scale foreshadowing of a wider disaster.

In the 1930s, a series of intense droughts struck the Plains states. Windstorms sucked great clouds of dry topsoil into the air and carried them away. This catastrophe, called the Dust Bowl, drove thousands of farm families off their land. Decades of drainage and wetland destruction had lowered water tables in the region by as much as fifty-nine feet, putting groundwater out of reach as an alternative supply.

The ecologist William Vogt wrote that wetland drainage plagued North America like a disease. In a 1938 booklet published by the Audubon Society, he laid out the multiple arguments for wetland conservation.[36] Wetlands capture rainwater, allowing it to percolate underground instead of rushing off the land carrying topsoil away with it. They nurture fish and wildlife, absorb excess water in times of flood, and act as reservoirs in time of drought.

Vogt's impassioned argument came in the midst of President Franklin Roosevelt's New Deal. To address mass unemployment and the farm crisis in the Plains, Roosevelt's administration created new federal agencies, among them the Soil Conservation Service and the Civilian Conservation Corps (CCC). The mission of the Soil Conservation Service was to encourage practices that would keep soil on the land. The agency promoted planting trees around fields as shelter belts, retiring marginal crop land, and planting legume cover crops in winter. It also became increasingly involved in the drainage business, seen as an essential part of good farming.

The CCC put thousands of young men to work—and much of the time, they worked on drainage. In 1939, the CCC redrained more than 1.5 million acres where ditches or tiles had grown clogged from neglect.[37] CCC workers also dug and repaired mosquito ditches on the Atlantic Coast and in Michigan, draining millions of acres by 1940.

During FDR's second term, the Works Progress Administration (WPA) continued large-scale drainage work, transforming thousands of acres of Louisiana wetland to farms and repairing and building more than twenty thousand miles of mosquito ditches. WPA workers built California's Shasta Dam, a project that contributed to drastic loss of marshes. By 1939, 85 percent of the Central Valley's wetlands were gone.

Meanwhile, the surviving wetlands of the Prairie Pothole region, in the northern Midwest, shriveled in the drought. Potholes are shallow marshes that were formed during the retreat of the glaciers at the end of the Ice Age. More than half of North America's waterfowl traditionally bred there. In 1934, the continent's waterfowl population reached a historic low of 27 million birds. Some species were in serious danger of extinction: Only fourteen whooping cranes survived.

Roosevelt instituted a President's Committee on Wildlife Restoration, charging its members to create a plan that would restore waterfowl habitat while helping to resolve the farm crisis.[38] On the committee were Thomas Beck, a leader of the duck hunter's group that would evolve to become Ducks Unlimited; the pioneering wildlife biologist Aldo Leopold, who'd witnessed firsthand the mass loss of pothole wetlands to drainage; and Jay "Ding" Darling, an editorial cartoonist who'd won two Pulitzer Prizes for his conservation-oriented drawings. The committee proposed that the government buy 4 million acres of marginal farmland to be restored as waterfowl nesting and breeding grounds, and more than 12 million acres total for wildlife of all kinds.

Soon after, Darling was appointed chief of the US Biological Survey. He enlisted CCC and WPA crews to build dikes to impound water for birds, reversing farm drainage systems to restore wetlands. Darling oversaw the purchase and re-engineering of drained land for refuges around the country. He protected the Aransas Refuge on the Texas coast, the last place where whooping cranes wintered in the US. Working with Finley, he raised funds to buy the rights to return water to the Malheur Refuge in eastern Oregon, which had been reduced to a dry alkali flat by water diversions.

While Darling and other conservationists struggled to protect and restore wetlands, other federal agencies continued to destroy them. The US Army Corps of Engineers was busily building flood control levees, wiping out wetlands in the process. The USDA subsidized farm drainage with millions of federal dollars.

In Darling's eyes, the USPHS was the worst offender. USPHS drained marshes nationwide in the name of malaria control. In 1936, the agency requested $74.5 million to drain coastal wetlands in Maine, New York, and other northern states that were malaria-free. The USPHS had refused to consult with Darling's Biological Survey on its drainage plans. Survey biologists reported cases of wetlands being destroyed under the guise of "malaria control" when the only benefit was to private landowners seeking more tillable acres on their property—in fact, the finished work left behind more mosquito breeding habitat than had been there before.[39] "The conclusion is inescapable," wrote Darling, "that the US Public Health Service under its existing administration is a pernicious racket."[40]

Between World War I and beginning of the New Deal, drainage work was seen as too expensive to be used to rid the South of malaria. Starting in 1935, the WPA provided labor for drainage projects on a massive scale: 544,414 acres of mosquito breeding sites drained through construction of 33,655 miles of ditches by 1942.[41] Soon after, malaria was gone from the American South.

It's difficult to pin down just how this happened. A growing contingent of biologists and conservationists were appalled by the WPA drainage program, which was carried out without regard to the biological intricacies of mosquitoes or malaria. Wetlands were drained without first testing for the presence of quad mosquito larvae. Melvin Goodwin, an ecologist studying mosquitoes in Georgia, often found quad larvae in malaria control ditches in greater abundance than in natural wetlands.[42] While some historians believe drainage played a pivotal role in extirpating malaria from the South,[43] others argue that the movement of impoverished tenant farmers out of the countryside and away from wet bottomlands during the Great Depression was key.[44]

In any case, malaria, along with the most deadly infectious diseases of the nineteenth century (cholera, tuberculosis, diphtheria, typhus, yellow fever), had been extirpated or brought under control in the US. Then, on December 7, 1941, the Japanese attacked Pearl Harbor, and the nation was drawn into World War II. Soldiers would be sent to Europe, Africa, and the South Pacific—places where the insect-borne diseases of malaria and typhus (a parasitic infection transmitted by body lice) were either endemic or exploding under the unsanitary conditions of battle. In the long history of war, these diseases had decided many military campaigns. US public health officials set out to make sure that insect-borne disease would not defeat the Allies.

To accomplish this, they focused on killing off the insect vectors. The weapon of choice had been pyrethrum, a compound derived from chrysanthemum petals, which had been used to fight body lice during the Napoleonic wars and was a popular insecticide in the US. Japan produced more than 90 percent of the chrysanthemums used to manufacture pyrethrum, however, and with the start of the war the supply of blossoms was cut off.

In the urgent search for a replacement insecticide, US researchers studied a new chemical being marketed by the Swiss manufacturer Geigy. In 1939, a Geigy scientist named Paul Müller had been assigned to find a chemical that would control invasive Colorado potato beetles. Müller dug up a formula first created in 1874, which synthesized dichloro-diphenyl-trichloroethane, soon to become known as DDT. The stuff made potato beetles vanish as if by magic. It was long-lasting too: In Müller's lab, a tube treated with DDT only once continued to kill flies for many weeks.

US researchers confirmed Geigy's claims: DDT powder dusted on humans did away with body lice. DDT sprayed from planes killed off mosquitoes (as well as many other insects). Some people exposed to the chemical got rashes, but there was no acute toxicity. Military researchers knew that DDT acted as a nerve toxin, and that heavy doses caused convulsions or death in lab animals.[45] In time of war, however, the short-term ability to fend off insect disease vectors outweighed any subtle questions of risk.[46]

In the fall of 1943, US forces pushed German occupiers out of Naples, Italy. The city had been torn apart by months of fighting and had no running water, gas, or electricity. In the first few months of American occupation, a growing number of typhus infections were recorded. By late December, the disease had become epidemic, and official reports estimated that more than 90 percent of the civilian population carried lice.

With the assistance of the Rockefeller Foundation, the US military began dusting every civilian with DDT powder. By March 1944, 1.3 million people had been dusted with DDT, and typhus had been vanquished in Naples. Though civilians had been given little choice in the matter, they were happy to be louse-free. "Neapolitans are now throwing DDT at brides instead of rice," claimed the *New York Times*.[47]

Through the rest of the war, the pesticide was widely used by the US military. A petroleum–DDT mixture was sprayed from planes to clear mosquitoes from beachheads before Allied invasions. An epidemic of dengue fever that broke out after the invasion of Saipan was quashed with a massive aerial spraying campaign. World War II became the first major war in which more soldiers died of combat than disease.[48] DDT was seen as a miraculous chemical that had been integral to the Allied victory.

Some of the largest chemical companies in the US mass-produced DDT for the war effort, and were soon marketing it to the public as a safe way to get rid of any and all insects. It was widely used against crop pests. In the postwar years, pesticide trucks rolled down suburban streets, emitting a fog of DDT spray that enveloped children playing outside.

There were some early warnings of the chemical's dangers. In August 1945, aerial spraying of DDT was tested on the New Jersey shore. Army officials reported that mosquitoes vanished, allowing troops to do their outdoor calisthenics unmolested. A representative of the State Agricultural Experiment Station in New Brunswick pointed out a downside: An experimental spraying on nine miles of beach caused a mass fish kill.[49] Researchers at the Patuxent Wildlife Refuge tested the effects of DDT on wildlife and began to find serious impacts on wild birds.[50] Biologist and science writer Rachel Carson tried to pitch an article on the dangers of DDT to *Reader's Digest* in 1945, but the editors weren't interested.

Agricultural pesticide use in the US skyrocketed after the war, climbing from 125 million pounds per year in 1945 to more than 600 million in 1955.[51] Industry developed new kinds of organochlorine pesticides. They were sprayed on the farms of the Midwestern corn belt and California's Central Valley. DDT and other organochlorines accumulated in the fatty tissue of exposed animals, becoming more concentrated as they moved up the food chain. The whole class of pesticides would prove especially deadly to predatory birds.

Malaria was gone from the US, aside from a few remnant pockets of infection in the South. Public health agencies continued to battle an already vanquished enemy on the theory that malaria would rise again if the fight was stopped. DDT spray campaigns replaced the ditching projects of the prewar era. The military agency Malaria Control in War Areas morphed into the Communicable Disease Center and continued pumping DDT into the environment. The agency used DDT sprayed on the inside walls of houses in the South, in addition to aerial spraying on wetlands. The modern incarnation of the Communicable Disease Center, the Centers for Disease Control and Prevention, still claims credit for knocking malaria out of the American South with DDT.[52] That notion is questionable. The single controlled study of the Communicable Disease Center's postwar spraying program found that the malaria parasite vanished from an untreated pond just as it did in a comparable pond sprayed with DDT.[53] Today the quad mosquito remains common in much of the US, though the malaria parasite is vanishingly rare.

In 1955, the World Health Organization established a program designed to rid the Earth of malaria using the sole tactic of killing mosquitoes with DDT. By 1969, anopheline mosquitoes had evolved resistance to the pesticide, and the effort was suspended.[54] More recent campaigns to control the disease in sub-Saharan Africa, where malaria remains endemic and up until a few years ago was killing millions of young children, have used pyrethrum sprayed on walls and insecticide-treated bed nets, with remarkable success. Cutting-edge research on malaria control now focuses on a range of creative tactics, from luring mosquitoes into traps using a chemical mix that mimics the smell of human sweat and breath, to bioengineering malaria-resistant mosquitoes.[55]

In 1957, a mosquito control plane flew over Duxbury, Massachusetts. It returned to spray repeatedly over the marshy property of Olga Owens Huckins, a friend of Rachel Carson's. The next day Huckins found several songbirds lying dead on the ground. "They were birds that had lived close to us, trusted us, and built their nests in our trees year after year," she wrote. "All of them died horribly. Their bills were gaping open, and their splayed claws were drawn up to their breasts in agony."

Huckins' outrage would inspire Carson to write her influential book, *Silent Spring*, in which she meticulously documented the ecological damage caused by organochlorine pesticides. Among the many examples Carson cited were a mass fish kill that followed the 1955 spraying of two thousand acres of Florida salt marsh with dieldrin, an organochlorine forty times more toxic than DDT. Crabs fed upon the many fish carcasses and the next day were dead themselves.[56]

Wild birds were also dying of pesticide poisoning at the Klamath Wildlife Refuge. Water was finally returned to the Lower Klamath refuge in 1942—not to recreate lost wildlife habitat, but because farmers in the dried-out bed of neighboring Tule Lake were being swamped in their own runoff. Irrigation water ran off cropland into the Tule sump, a remnant of the original lake. As the number of homesteaders grew, so did the volume of runoff and the water level in the sump. In 1928, the Tule sump was declared a bird refuge. Later, at William Finley's urging, President Franklin Roosevelt would triple the refuge's size to 37,000 acres. Waterfowl flocked there.

Eventually, the Tule sump would threaten to flood nearby farms. Meanwhile, peat fires and dust storms continued to plague the desiccated bed of Lower Klamath Lake. A solution was created by a Reclamation Service engineer, J.R. Iakisch. He envisioned a six-thousand-foot tunnel dug through the ridge separating the two lake beds so that excess water from the Tule sump could be pumped to the dry bed of the Lower Klamath.[57] In 1940, the Klamath Drainage District, representing private landowners, signed a cooperative agreement with the US Biological Survey. The government agreed to build the drainage tunnel and to build a dike across the southern edge of Lower Klamath Lake to prevent flooding of neighboring farms. The long-lost wetlands of the Lower Klamath were resurrected, at least in part. They were fed with agricultural runoff, and the fate of the farms created by the Reclamation Service dams was forever tied to that of the area's wildlife refuges.

From a bird's point of view, the Klamath Basin refuges were irresistible—even if the waters were tainted with alkali, fertilizer, and pesticides. At the peak of fall migration, 7 to 8 million ducks, geese, and swans gathered there. "There is probably no more important waterfowl area in the country," Interior Secretary Stewart Udall told a Congressional hearing in 1962. The Klamath refuges, he said, act like the waist of an hourglass, through which all the birds of the Pacific Flyway funnel on their annual migrations.[58]

Irrigation water flowing to Klamath Basin farms was recycled seven times before it ran to the wildlife refuges. By then it was tainted with the chemical residues of postwar agriculture. Between 1960 and 1964, hundreds of white pelicans, egrets, and western grebes died at the Klamath Basin refuges. A few birds were seen to drop from a flock in flight, dead before they hit the water. Adults and chicks in breeding colonies suffered tremors and convulsions as they died. The affected species were all fish-eating birds, vulnerable to the impacts of organochlorine pesticides because the chemicals concentrate in their prey. A 1966 study found that birds from the Tule Lake and Lower Klamath refuges carried high concentrations of DDT and the organochlorines toxaphene and endrin in their tissues.[59] Previous studies had shown that small concentrations of organochlorines in water could translate into deadly doses for fish-eating birds.

There were also subtler effects. DDT poisoning caused female birds to produce thin-shelled eggs that broke before the embryo inside could mature. The syndrome was observed in many species of North American birds after 1946 when widespread DDT use began.[60] Eggshell thinning contributed to major declines in the populations of pelicans, bald eagles, and peregrine falcons.

Agricultural use of endrin was banned in 1964. In 1972, after a long battle by conservationists, DDT was banned in the US. Because organochlorines persist in the environment, white pelicans breeding at the Klamath refuges were producing abnormally thin-shelled eggs as late as the 1980s.[61]

In the midst of the mass drainage and poisoning of the mid-twentieth century, a new generation of wetland defenders emerged. In the 1950s, pioneering work by ecologist Eugene Odum at Sapelo Island, Georgia, showed that salt marshes are among the most biologically productive ecosystems on the planet, and provide nutrients vital to the growth of fish and shellfish in nearby marine waters. Representative Henry Reuss of Wisconsin successfully sponsored a 1962 law that forbade the USDA from helping farmers in the Prairie Pothole region to drain marshes if wildlife was at risk. Activists fought to stop the US Army Corps of Engineers from destroying the last remnants of Pennsylvania's Tinicum Marsh.[62]

When Arcata's sewage rebels stumbled onto the idea of a treatment marsh, the idea of wetland restoration was still a low murmur in the ambient roar of destruction. Building the town's marsh would take not just bold scientific invention but political cunning as well.

NOTES

1. Vileisis, A. (1997). "Discovering the unknown landscape: a history of America's wetlands," p. 42. *Island Press, Washington DC and Covelo, CA.*
2. Ibid.
3. Ibid.
4. Brush, G. (2009). "Historical land use, nitrogen, and coastal eutrophication: a paleoecological perspective." *Estuaries and Coasts* **32**(1): 18–28.
5. Dobson, M. (1980). "Marsh fever: geography of malaria in England." *Journal of Historical Geography* **6**(4): 357–389.
6. Bindewald, J. "Death and mortality in the Colonial Chesapeake." http://public.gettysburg.edu/~tshannon/341/sites/Death%20and%20Mourning/chesapeake.htm.
7. Ibid., p. 32.
8. Brown, S.R. (1814). "Views of the campaigns of the north-western army, &c. comprising sketches of the campaigns of Generals Hull and Harrison, a minute and interesting account of the naval conflict on Lake Erie, military anecdotes, abuses in the army, plan of a military settlement, view of the lake coast from Sandusky to Detroit." *Samuel Mills, Burlington, Vermont.*
9. Brooks, M. (2006). "A pocket of pestilence: a historical and epidemiological evaluation of Northwest Ohio's nineteenth-century reputation as an unhealthy region." *Master's thesis, University of Toledo.*
10. Ibid.
11. Everett, H. (1882). "History of Sandusky County, Ohio." *H.Z. Williams & Bro.* archive.org.
12. Ibid., p. 206.
13. McCorvie, M.R., Christopher L. Lant (1993). "Drainage district formation and the loss of midwestern wetlands, 1850–1930." *Agricultural History* **67**(4): 13–39.

14. Wilhelm, P.W. (1984). "Draining the black swamp: Henry and Wood Counties, Ohio, 1870–1920." *Northwest Ohio Quarterly* **56**(3): 79–95.
15. Ibid.
16. Kaatz, M.R. (1955). "The Black Swamp: a study in historical geography." *Annals of the Association of American Geographers* **XLV**(1): 1–36.
17. Wilhelm, P.W. (1984). "Draining the black swamp: Henry and Wood Counties, Ohio, 1870–1920." *Northwest Ohio Quarterly* **56**(3): 79–95.
18. Biebighauser, T.R. (2007). "Wetland drainage, restoration, and repair," p. 5. *University Press of Kentucky, Lexington, Kentucky.*
19. Waring, G.E. (1902). "Draining for profit and draining for health," p. 7. *Orange Judd Company, New York.*
20. Cassedy, J. (1962). "The flamboyant Colonel Waring: an anticontagionist holds the American stage in the age of Pasteur and Koch." *Bulletin of the History of Medicine* **36**: 163–176.
21. PBS (2006). "The Great Fever." *pbs.org.*
22. Cassedy, J. (1962). "The flamboyant Colonel Waring: an anticontagionist holds the American stage in the age of Pasteur and Koch." *Bulletin of the History of Medicine* **36**: 163–176.
23. PBS (2006). "The Great Fever." *pbs.org.*
24. Cassedy, J. (1962). "The flamboyant Colonel Waring: an anticontagionist holds the American stage in the age of Pasteur and Koch." *Bulletin of the History of Medicine* **36**: 163–176.
25. Patterson, R. (1989). "Dr. William Gorgas and his war with the mosquito." *Canadian Medical Association Journal* **141**: 596–599.
26. LePrince, J.A., A.J. Orenstein, L.O. Howard (1916). "Mosquito control in Panama." *G.P. Putnam's Sons, New York and London, The Knickerbocker Press.*
27. LaPointe, P. (1987). "Joseph Augustin LePrince: his battle against mosquitoes and malaria." *West Tennesee Historical Society Papers* **41**: 48–61.
28. Vileisis, A. (1997). "Discovering the unknown landscape: a history of America's wetlands," p. 113. *Island Press, Washington DC and Covelo, CA.*
29. Vogt, W. (2013). "Thirst on the land: a plea for water conservation for the benefit of man and wild life." *National Association of Audubon Societies* **Circular 32** (Literary Licensing, LLC).
30. Akerknecht, E.H. (1945). "Malaria in the Upper Mississippi Valley 1760–1900," p. 129. *Bulletin of the History of Medicine* **Supplements**(4): Johns Hopkins Press, Baltimore, MD.
31. Finley, W. (1907). "Among the pelicans." *The Condor* **9**(2): 35–41.
32. Mathewson, W. (1986). "William L. Finley: pioneer wildlife photographer," p. 57. *Oregon State University Press, Corvallis.*
33. Foster, D. (2002). "Refuges and reclamation: conflicts in the Klamath Basin, 1904–1964." *Oregon Historical Quarterly* **103**(2): 150–187.
34. Ibid.
35. Ibid.
36. Vogt, W. (2013). "Thirst on the land: a plea for water conservation for the benefit of man and wild life." *National Association of Audubon Societies* **Circular 32** (Literary Licensing, LLC).
37. Vileisis, A. (1997). "Discovering the unknown landscape: a history of America's wetlands." *Island Press, Washington DC and Covelo, CA.*

38. Garone, P. (2011). "The fall and rise of the wetlands of California's great Central Valley," p. 137. *University of California Press, Berkeley, Los Angeles, London* (HSU Library).
39. Vogt, W. (2013). "Thirst on the land: a plea for water conservation for the benefit of man and wild life." *National Association of Audubon Societies* **Circular 32** (Literary Licensing, LLC).
40. Darling, J. (1936). "US Public Health Service vs. conservation." *Outdoor America* **1**(10): 3.
41. Sledge, D., George Mohler (2013). "Eliminating malaria in the American South." *American Journal of Public Health* **103**(8): 1381–1392.
42. Way, A.G. (2015). "The invisible and indeterminable value of ecology: from malaria control to ecological research in the American south." *Isis* **106**(2): 310–336.
43. Sledge, D., George Mohler (2013). "Eliminating malaria in the American South." *American Journal of Public Health* **103**(8): 1381–1392.
44. Humphreys, M. (2014). "Malaria in America." In Snowden, Frank and Richard Bucala, editors. "The global challenge of malaria." World Scientific.
45. Simmons, J.S. (1945). "How magic is DDT?" *Saturday Evening Post* **217**(28): 18–86.
46. Kinkela, D. (2011). "DDT and the American century," p. 21. *University of North Carolina Press, Chapel Hill.*
47. Ibid., p. 29.
48. PBS (2010). "Rachel Carson and *Silent Spring*." *American Experience.* http://www.pbs.org/video/2365077442/.
49. Anonymous (August 9, 1945). "Fish killed by DDT in mosquito tests." *New York Times.*
50. Robbins, C., Robert Stewart (1949). "Effects of DDT on bird population of scrub forest." *Journal of Wildlife Management* **13**(1): 11–16.
51. PBS (2010). "Rachel Carson and *Silent Spring*." *American Experience.* http://www.pbs.org/video/2365077442/.
52. CDC. "Elimination of malaria in the United States 1947–1951." https://www.cdc.gov/malaria/about/history/elimination_us.html.
53. Slater, L., Margaret Humphreys (2008). "Parasites and progress: ethical decision-making and the Santee-Cooper malaria study, 1944–49." *Perspectives in Biology and Medicine* **51**: 103–120.
54. Kinkela, D. (2011). "DDT and the American century," p. 1. *University of North Carolina Press, Chapel Hill.*
55. Levy, S. (2016). "New hope for malaria control." *BioScience* **66**(6): 439–445.
56. Carson, R. (1962). "Silent Spring." *Houghton Mifflin, Boston and New York.*
57. Foster, D. (2002). "Refuges and reclamation: conflicts in the Klamath Basin, 1904–1964." *Oregon Historical Quarterly* **103**(2): 150–187.
58. Ibid.
59. Keith, J.O. (June 1966). "Insecticide contaminations in wetland habitats and their effects on fish-eating birds." *Journal of Applied Ecology* **3**(Supplement: Pesticides in the environment and their effects on wildlife): 71–85.
60. Boellstorff, D., Harry Ohlendorf, Daniel Anderson, et al. (1985). "Organochlorine chemical residues in white pelicans and western grebes from the Klamath Basin, California." *Archives of Enivronmental Contamination and Toxicology* **14**: 485–493.
61. Ibid.
62. McCormick, J. (1971). "Tinicum Marsh." *Bulletin of the Ecological Society of America* **52**(3): 8, 23.

8

Revolution

The fight over the Humboldt Bay Wastewater Authority (HBWA) project had turned bitter and personal. HBWA's attorney, John Stokes, and most of its board members had lobbied hard against Arcata's alternative treatment plan. Dan Hauser, usually diplomatic, seethed with resentment.

"HBWA has set itself up as the enemy," he wrote in a September 1977 opinion piece in the Arcata *Union*. "Therefore, we have no alternative but to defend ourselves by attacking HBWA . . . We must stop this $52 million boondoggle."[1] Hauser, still Arcata's representative on the HBWA board, pledged to work toward the "total redesign or total destruction" of the regional sewage system.

Other members of HBWA were growing panicky. The Committee for a Sewer Referendum's lawsuit kept the board from issuing bonds to finance construction, while inflation caused the project's already huge price tag to balloon. Concealing the move from Hauser, the board applied for a $5.9 million loan from the US Environmental Protection Agency (EPA). Arcata, at Mayor Hauser's suggestion, promptly sued HBWA for seeking the loan without the city's consent.

Meanwhile, Hauser organized an appeal for Arcata's wetland treatment system before the State Water Resources Control Board. The city mustered support from representatives of the US Fish and Wildlife Service and Audubon Society, along with academic experts on Humboldt Bay oysters and low-tech sewage treatment. Wade Rose, the shaggy upstart from the governor's Office of Appropriate Technology, would speak. After Stokes cross-examined Rose at the regional board hearing, "it became a crusade for the entire Office of Appropriate Technology," Hauser explains. "They singled out HBWA as the ultimate in obsolete technology and concrete overkill."

When the Arcata contingent arrived at the state board hearing in Sacramento, one of the board members, brandishing a newspaper clipping in his hand, called Hauser forward. The clipping was a story from the Arcata *Union*, quoting Hauser saying that the marsh project would not get a fair hearing. "He asked why I was there if I believed they were already biased against me," Hauser remembers. "I told him we have to go through this process to get to the next step."

The state board heard testimony on Arcata's vision of wetlands treatment for two days. The board refused to reconsider the Bays and Estuaries policy, the driving force behind the regional sewage project. It declined to give Arcata an exemption that would release the city from the HBWA agreement. For the first time, however, the board recognized some value in the idea of constructed wetlands for treating wastewater. State board staff suggested Arcata be given a grant to plan a pilot marsh treatment project. In another conciliatory gesture, they suggested that HBWA delay construction of the controversial east bay sewage interceptor.

During a recess, Hauser talked with Don Maughan, chair of the state board, about the possibility of full funding for a pilot marsh project. Maughan saw this as a compromise, giving Arcata a chance to prove out its marsh concept while the regional sewage system moved forward. A pilot study of treatment wetlands would take several years, after which Arcata might be able to withdraw from HBWA—having paid millions for the project's construction.

Hauser had other ideas. He was enthusiastic about a state-funded pilot marsh, which he saw as a wedge that could ultimately be used to topple HBWA. Gearheart was confident the marshes would work to make the city's sewage effluent cleaner, if given a chance. Hauser intended to use state funding to launch the pilot while he and other activists held the HBWA project off. In time, he believed the behemoth project would crumble under the weight of local resistance.

"My opposition—and the Arcata city council's opposition—to HBWA will not end," Hauser said.[2]

A new player in the anti-HBWA contingent was Bill Bertain, a young attorney practicing in Eureka. Short, stocky, and intense, Bertain had been born and raised in the Humboldt County lumber town of Scotia. He'd served in the military before attending law school, and was admitted to the California bar in 1976. He found a slot at a law firm led by Bob Dunaway, a descendant of a long-time Humboldt family. Bertain would pay a high price for his passionate involvement in the fight against the HBWA project, losing the first two jobs in his legal career in the course of two years. "I like a fight," says Bertain, "and I cut my teeth on the HBWA battle. The mandate from the state didn't make sense. Bigger isn't always better, and the experts don't always know best."

Bertain first learned the details of the regional sewage plan from Jackie Kasun. Both were devout Catholics and anti-abortion activists. He hit it off with Dan Ihara, Kasun's cofounder at the Committee for a Sewer Referendum. HBWA's refusal to allow Humboldt citizens to vote on the project infuriated Bertain. The engineers and officials behind the regional sewage plan, he felt, suffered from "arrogance run amok."

A new strategy against HBWA emerged when, at his first legal firm, Bertain was assigned to represent the business interests of Lloyd Hecathorn. For years, the only market for redwood chips from local timber companies had been the Simpson and Georgia-Pacific pulp mills on the Samoa Peninsula—and the corporations controlled the price of wood chips. Hecathorn created North Coast Exports, a company that bought wood chips from several small Humboldt timber companies, paying a more generous price than the local pulp mills did, and shipped them to markets in Japan. On the day North Coast Exports began operation, says Bertain, Simpson and Georgia-Pacific were forced to triple their pay rate for wood chips.

Hecathorn's business relied on big ships moving in and out of Humboldt Bay, and he was appalled by the HBWA design, which called for a cross-bay pipeline carrying raw sewage from Eureka to the planned activated sludge plant on the Samoa Peninsula. Cargo ships had to drag their massive anchors in order to turn around, and an anchor might rupture the pipe, spewing raw sewage into the heart of the bay. Arthur Einerfeld, a consulting engineer who had studied Humboldt's sewage problems, had warned of this risk in a 1974 report. HBWA's general manager, John Stratford, dismissed this concern as "political." The project design, including the trans-bay sewer line, had already been approved by the coastal commission, the harbor district, and the US Army Corps of Engineers.

Hecathorn also had a more self-interested gripe: HBWA planned to build its treatment plant on a parcel between the two pulp mills on the Samoa Peninsula. He had his eye on that same piece of land for a business project, a power plant that would generate energy by burning scrap wood. Using the last large chunk of industrial-zoned real estate on the bay for a sewage plant would be "a disaster for the economic development of the community," Hecathorn claimed.

"The chances of our backing up from that site," Stratford replied, "are zilch."[3]

Hecathorn formed a group he called Concerned Citizens for the Development of Humboldt Bay, with Bertain as his legal advisor. Hecathorn said the cross-bay sewer line would destroy commerce on the bay. Ship pilots would fear becoming liable if they accidentally caused a break in the pipe, and would refuse to work near it. A group of bar pilots, responsible for guiding ships through the treacherous bay entrance, agreed and joined in resisting HBWA. The idea of piping raw sewage under the bay, Bertain says, "was so stupid it was like a gift to the opposition."

In September 1977, Bertain sued HBWA on behalf of Concerned Citizens, alleging that the cross-bay sewer pipeline would represent a serious health hazard. He also filed lawsuits against the Humboldt Bay Harbor District and the California Coastal Commission for issuing HBWA permits to construct the pipeline. He found a geomorphologist who warned that the proposed sewer pipe would cross a fault line, making it susceptible to destruction in an earthquake.

Bertain couldn't stop thinking about ways to combat HBWA. He worked long hours, researching the authority's legal weak spots, chain-smoking in his office, sleeping little. The first law firm he'd worked at let him go because he wasn't bringing in enough money—anti-HBWA research didn't make for billable hours. Bertain moved to the conservative firm of Falk, Buxton and Brown, bringing his work for North Coast Exports and Concerned Citizens with him. Six months later, Bertain came across an EPA audit of the HBWA project, which pointed out a conflict of interest: Robert Kelly, of Winzler & Kelly Engineers, had served as interim manager for HBWA, stepping down only two weeks before his firm was chosen as the primary contractor on the lucrative project. The audit also mentioned some questionable use of grant monies.

Bertain's boss, Charles Buxton, had social connections with Winzler & Kelly, and a strong distaste for controversy. Nonetheless, Bertain passed the audit information to Dan Ihara, who took it to the press. Winzler & Kelly sued Ihara for defamation. (The suit was dropped, perhaps because the spectacle of deep-pockets engineering consultants going after an impoverished activist made for yet more negative publicity.) Bertain lost his job, but kept fighting HBWA. At a meeting one day, John Winzler, Kelly's partner, told Bertain that anyone who thought they could force a change in state water policy was dead wrong and pathetically naïve.

"I told him," remembers Bertain, "that if democracy retains any meaning, then people can at least attempt to change stupid governmental policies. That they should be changed if they are bad for the commonweal, and I thought we could make it happen."

Bertain set up his own office in Eureka. He's been practicing law there ever since.

Stokes and Stratford flew to Washington, D.C. to plead with EPA officials for approval of the contested loan to HBWA. "We made no bones about the Arcata thing," Stratford said. "We told them that Arcata's been using the wastewater authority as a protective blanket while they do their best to screw everyone else in the area."

EPA denied the loan. Construction on the regional sewage project was stymied: There was no money to pay contractors. The Committee for a Sewer Referendum was still pursuing a suit against the wastewater authority, and no bonds could be issued until the case was resolved.

Dan Ihara had been fighting the HBWA project for a year. His opponent had the resources to invest many thousands of dollars in legal defense. For the grassroots Committee for a Sewer Referendum, keeping the lawsuit alive was a constant struggle. The organization's funds were depleted, and Larry Eitzen, the local attorney who was handling the case on a pro bono basis, had to take some time for paying work.

One evening Ihara was at home, watching his young daughter, and wondering if taking on the HBWA battle had been a colossal mistake. The phone rang, the caller a student of Bob Gearheart's who had come into a small inheritance and wanted to donate $800 to the Committee for a Sewer Referendum. While Ihara got wrapped up in an intense conversation about the HBWA fight, his daughter covered the entire kitchen with a coat of red tempera paint. The spectacular mess was worth it. "I couldn't let the chance slip away," Ihara remembered. "That $800 turned out to be crucial."

By the end of 1977, thirty-one of George Allen's salmon, marked by a distinctive notch cut into their fins, had returned to Arcata's creeks as spawning adults (Fig. 8.1). Gearheart, Klopp, and Allen had written up a proposal for a pilot marsh, which they hoped would bring funding for construction of a series of sizeable wetlands adjacent to Mt. Trashmore, capable of filtering 100 percent of the effluent from the city's oxidation ponds. Through Christmas and the New Year's holiday, the city's sewage activists had an interlude of peace and optimism.

Then, in January 1978, the regional board struck back, announcing that it would impose a building moratorium on Arcata. No new sewer hookups would be allowed until the city had ended all sewage discharges to Humboldt Bay. There was an exception for building projects that already held city permits. When Hauser got word of the board's intentions, city staff began cranking out building permits

Figure 8.1 George Allen with one of the first salmon raised at his Arcata wastewater aquaculture project to return as an adult, 1977. Photo originally published in *Arcata Union.*

at high speed. Three million dollars' worth of construction permits—representing six months of normal growth in Arcata—were processed in two weeks.

"We aren't going to sell out the citizens of Arcata and the environment of Humboldt Bay because of these threats," vowed Wes Chesbro, who by then held a seat on the regional board's policy advisory committee, in addition to his membership on the city council. "The board is so inflexible and backward thinking that it punishes communities for seeking innovative solutions."

David Joseph and other regional board representatives claimed that Arcata's pilot marsh would never be funded, that if built it could never succeed. The city's wetland dreams were irrelevant. The HBWA project was what mattered, and it must go forward.

Officials at the state board made it clear that a pilot project would have to be a small, controlled experiment, receiving no more than 10 percent of the effluent flowing from the city's oxidation ponds. The pilot should include a series of small marshes where water purification under different flow rates and with different species of aquatic plants could be tested. "We also expect that the HBWA regional project will proceed as planned," said Curtis Swanson, chief engineer with the state board.[4] Arcata's marsh treatment, if it could be proven to work, would be delayed by bureaucratic proceedings for years. It might eventually be used to increase treatment capacity as the city grew, but the offer to fund a pilot project was decidedly not meant to give the city an escape from HBWA.

Hauser was undeterred. If the state wouldn't fund a full-scale wetland, he'd find the money somewhere else. "One way or another," he said, "we're going to get this project built."

The HBWA board began to show the first signs of compromise in April 1978. Bob Brown, the Humboldt Community Service District's representative on the board, presented a scaled-down version of the regional project. The east bay interceptor running from Arcata to Eureka, the original source of Hauser's opposition, would be dropped. Instead a sewer line would run from Arcata to the planned central treatment plant along the Samoa Peninsula, on the western edge of the bay, where development was unlikely. The treatment plant itself would use trickling filters instead of activated sludge, lowering energy and construction costs dramatically. A pipe carrying raw sewage from Eureka to Samoa would still be planted under the bay floor, but would be placed five feet deeper to protect against breakage from ship anchors (Fig. 8.2).

All these changes had been suggested over and over by various HBWA opponents, but the projects' engineers had refused to budge on any point. They'd insisted that only activated sludge, the most expensive and energy-intensive kind of treatment, would meet federal regulations. After reading through new state and federal rules adopted following the 1977 revision of the Clean Water Act, Brown concluded that was not the case.

Ed Estes, McKinleyville's representative to HBWA, still saw Arcata's attempt to bolt from the HBWA project as a sign of moral weakness. Estes liked the compromise proposal, though, and so did Hauser (Fig. 8.3). Dropping the east bay interceptor would save both towns hundreds of thousands of dollars. It was the first time they'd agreed on anything in months. Other opponents were unsatisfied: Jackie Kasun, the economist, could see no justification for even a scaled-down version of the regional project, and Bertain didn't want raw sewage piped under the bay under any circumstance. The tide of resistance kept rising, and state officials continued to claim it didn't matter.

The Eureka Chamber of Commerce passed a resolution opposing the Bays and Estuaries policy. So did the local harbor commission, and then the Humboldt County Board of Supervisors. The county's political balance had tipped when two anti-HBWA candidates, Danny Walsh and Eric Hedlund, were elected as supervisors. Urged on by hot-tempered Danny Walsh, the supervisors wrote to the district's state senator and assemblyman, asking them to introduce legislation rescinding the state ban on discharges of treated sewage to bays.

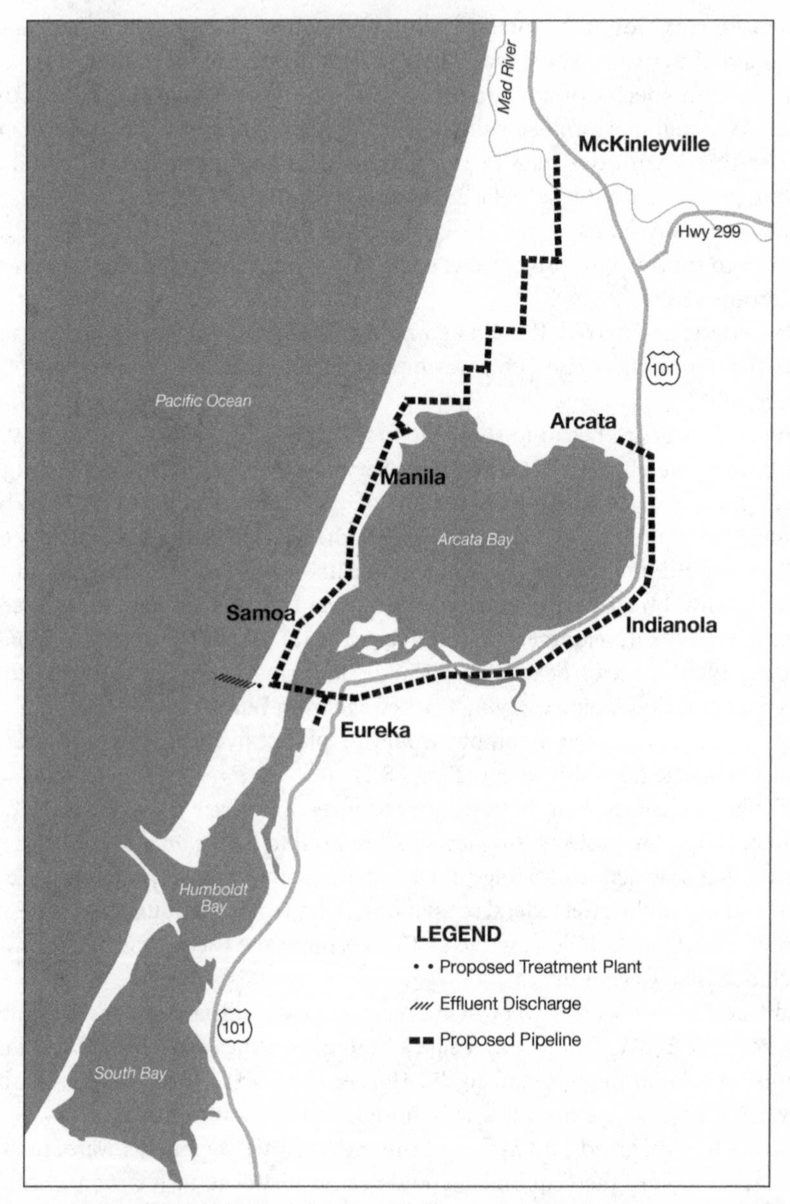

Figure 8.2 Map of Humboldt Bay showing the proposed regional sewage project to be built by HBWA, circa 1977. Graphic by Leslie Scopes Anderson.

Walsh became the county's representative on the HBWA board, a change that transformed board meetings into shouting matches and raised the collective blood pressure of the four board members who were struggling to get the regional system built. Fed up with all the fuss, David Joseph hit the entire Humboldt Bay region with a building moratorium.

"Dr. Joseph has been a black plague on this county for years and years now," Walsh commented, with his trademark lack of diplomacy. "I've really had it with that guy."

Walsh felt the opposition coming together. "It's steamrolling now," he told a reporter.[5]

Gearheart and Hauser traveled to Sacramento in June 1978 to discuss plans for the pilot marsh study with representatives of the state board. The meeting was amicable until Hauser asked if the building moratorium in Arcata could be lifted, given the progress toward wetlands treatment. The answer was no—not until the HBWA project was completed and Arcata was hooked up to it. Days later, Arcatans voted on an advisory referendum meant to gauge citizen support for the city council's risky wastewater fight: 76 percent supported the struggle for alternative sewage treatment.

At the annual conference of the League of California Cities, Chesbro introduced a resolution calling on the state to stop promoting regional sewage systems and to change the Bays and Estuaries policy. It passed by an overwhelming majority. Soon after, Arcata applied to the California Coastal Conservancy for a grant to build a series of freshwater wetlands near Mt. Trashmore, which would create important wildlife habitat on what had for years been a blighted waterfront. Hauser was working toward his promise to get a full-scale marsh built with or without the support of the state board.

In January 1979, Bertain, having been fired for a second time for his outspoken anti-HBWA activism, opened his own law office in Eureka. Now he was free to engage in outright battle against the regional sewage project—and he'd been thinking for months about the way to win.

The Committee for a Sewer Referendum had lost the final appeal in its lawsuit. HBWA could at last issue bonds, as soon as it received a new permit from the harbor commission. Chesbro agreed to stand as plaintiff in a suit filed by Bertain, demanding that HBWA file a more detailed Environmental Impact Report on the cross-bay pipeline. The Harbor Commission then issued HBWA a permit, which would become effective only when the new lawsuit was resolved. Once again, the opposition had halted construction.

The wastewater authority reacted by voting, despite protests from Hauser and Walsh, to demand a tenfold increase in funding from member entities, including Arcata, Eureka, McKinleyville, and the county. (Inflation had driven the project cost up to $68 million by March 1979.) Hauser promptly sued HBWA for this move, which he said violated the joint powers agreement that had created the wastewater authority in 1975.

Hauser worked as a very public double agent. He'd attend HBWA meetings where the lawsuits Arcata and Chesbro had brought were discussed; as a board member, he could not be excluded. Then he'd make the short drive to Bertain's office, which Hauser called "combat central." There the opposition gathered to plot strategy.

Lawsuits, Bertain explained to his comrades, were just one part of a three-pronged attack. Initiative petitions should be circulated in every community that was part of HBWA, demanding a vote on the future of the regional sewage system. If even one town voted against the project, HBWA would break apart.

The third tactic was to get the district's state legislators—Senator Barry Keene and Assemblyman Doug Bosco—to push for change in the Bays and Estuaries policy. Hauser, Walsh, and Arcata city attorney John Corbett were among a Humboldt delegation that traveled to Sacramento to make their case to Bosco.

Corbett drafted a bill altering the Bays and Estuaries policy. Soon after, Bosco announced that new legislation would be introduced in both houses, allowing discharges of treated sewage to Humboldt Bay if they could be shown to cause no damage to the ecosystem.[6]

Hauser and Corbett attended the hearing on the bill, held by the state assembly's Water, Parks and Wildlife Committee. "I'd never been to a legislative hearing in my life," Hauser remembers. He walked in and sat next to Bosco, trying to ignore the nervous flutter in his gut.

Bosco asked for a pitcher of water and poured some into a glass. "Under the current interpretation of the Bays and Estuaries policy," he explained, "you can't discharge this into Humboldt Bay." Then he drank the water down.

Bosco's illustration of the unreasonable standards of the state and regional boards grabbed the attention of everyone in the hearing room. The only exception allowed under the Bays and Estuaries policy, he explained, was in the case of a sewage discharger proving enhancement of the bay—but no one had defined what was meant by enhancement.

Don Maughan, chair of the state board, was in the room. Hauser remembers Maughan as "a very courtly gentleman, confined to a wheelchair. He was off to one side in his chair, and had a couple of aides with him."

Figure 8.3 Dan Hauser near the end of his term as Arcata's mayor, 1982. Photo courtesy Dan Hauser.

An assemblyman turned to Maughan and asked if he could explain what would qualify as "enhancement of receiving waters." In calm tones, Maughan replied that he wasn't sure of the definition, but he could send one of his aides down to the office to look it up.

There was a short, stunned silence. If the chair of the state board didn't know what the policy meant, how could a town like Arcata hope to meet the standard?

"The committee approved the bill unanimously in about thirty seconds," says Hauser. "I never opened my mouth." He suspected that Maughan had played dumb out of sympathy with Arcata's cause.

Hauser and Corbett passed HBWA's flustered representatives on the steps of the state Capitol. Their chartered plane had been delayed due to fog at the Arcata airport. In the thirty minutes they spent on the runway, they'd lost an important battle.

Assemblyman Bosco informed the state board that if it approved Arcata's marsh alternative within six months, he would not push the legislation forward. "That bill," says Corbett, "was a sledgehammer." The threat of legislation got the state board's immediate attention. The board announced a "fact finding hearing" to take place in Eureka in April.

Meantime, meetings of the HBWA board, which had become a quirky form of political theater, continued. Hauser and Walsh, the two anti-HBWA members, were vocal and were almost always voted down. "Last week's meeting," reported the *Arcata Union*, "featured board member attacking board member (verbally), grumbling about press coverage, impassioned oratory from the audience, the usual confusion over which side was voting which way, and not inconsiderable laughter." Ernie Cobine, a staunchly pro-HBWA board member from Eureka, pounded his shoe on the table, Khrushchev-style, to restore order.[7]

A relaxation of the Bays and Estuaries policy would reduce the cost of the regional sewage project by millions of dollars, and Hauser asked the board to take a formal stand in favor of such a change. The majority of the HBWA board, however, believed that in signing on to the joint powers agreement that created the authority, each community had essentially taken an oath to follow the dictates of the project's consulting engineers and the regional board. John Stokes, HBWA's attorney, repeated his legal opinion that the board's taking any stand regarding the Bays and Estuaries policy would be a violation of the joint powers agreement.

The upshot of all this was that Arcata had to bear the expense of hiring expert witnesses to testify at the upcoming state board hearing, at a cost of about $20,000. The city had already invested so heavily in its sewage battle that nobody questioned this. In the week before the state board hearing, the California Coastal Conservancy announced it would fund construction of three large freshwater marshes on the city's waterfront—a project that would go forward regardless of the outcome of the wastewater war. This was a tremendous morale boost for Hauser and his Arcata co-conspirators.

The state board hearing, held at the Elks Lodge in Eureka, unfolded over two long days in April 1979. Day one was a clash between experts. Roger Johnson, an engineer with the state board, laid out the history of the problem as his agency saw it. A 1971 study by the consulting engineers Baruth & Yoder had concluded, based on a mathematical model of Humboldt Bay, that sewage discharges contributed to contamination of oyster beds. They also predicted that within five to ten years the shallow waters off Arcata would turn anoxic and cease to support normal aquatic life. In 1974, the state had implemented its Bays and Estuaries policy, banning wastewater discharges to most of California's enclosed bays. This decision was based on the dire problems in heavily polluted East Coast estuaries, including Long Island Sound and Chesapeake Bay, and on the decline of San Diego Bay and its dramatic revival after urban sewage was piped out to the Pacific. Johnson summed up the stance of the state board's staff: A regional sewage project with ocean discharge was the only answer.

Hauser argued that the entire state strategy for Humboldt Bay was based on the Baruth & Yoder prediction that sewage discharges would choke the life out of the bay—a prediction that was in error. "We intend," he said, "to establish that Humboldt Bay is alive and healthy, that properly treated effluent discharge is beneficial to the bay, that the HBWA project is a waste of money that won't achieve its purpose."

Hauser was followed by a parade of professors from Humboldt State University, who had conducted the first substantive studies of the bay's ecology. The populations of seaweed, benthic creatures, and plankton were all diverse and healthy—comparable to those found in near-pristine Tomales Bay on the Marin County coast. Zooplankton typical of the open ocean were found off Arcata, a sign that ocean waters flushed far into the bay. That notion had been confirmed by tracking the movement of floating markers, called drogues, on outgoing and incoming tides. Concentrations of dissolved oxygen in the bay were high—sometimes higher than levels in the open ocean. "The bay will never have a problem with oxygen," testified oceanographer John Pequegnat, "because of the shallow nature of the bay and the turbulent mixing of bay waters."

Bob Gearheart explained that the HBWA project would do nothing to solve the problem of bacterial contamination of oyster beds, because the vast majority of the microbes were carried to the bay in stormwater runoff that ran, untreated, into the bay. Manure from cow pastures was a major source. Leaky sewer lines were also a problem, but the HBWA project made no provision for repair or upgrade of existing sewer lines.

The only real data the state had on the condition of Humboldt Bay came from Public Health Department reports on oyster contamination and a set of dye studies done by the regional board staff. They'd adapted the fluorescent dye technique David Joseph had used to demonstrate that streams were tainted during aerial spraying of herbicide on timber land. Dye released at the Arcata and Eureka sewage outfalls sloshed over to nearby oyster beds in a matter of hours, proving that wastewater discharges mixed with the water filtered by oysters. But that said nothing about the quality of the water, or the source of bacteria that contaminated shellfish during the heavy rains of winter.

Arcata's ecological arguments should have been convincing, but the Humboldt experts took it too far. They claimed that nutrients from treated sewage were beneficial to the bay, because they replaced the lost input of organic matter from the thousands of acres of salt marsh that had been destroyed since settlement in the 1850s. This was a very rough guess—one that sounded good to the Arcata contingent. It was, however, as unsubstantiated as the state's assumption that sewage was killing the bay.

Given the many well-known cases of ecological disaster fueled by nutrient overload in other estuaries, the notion that treated sewage was healthy for Humboldt Bay was a very hard sell. The state board members were skeptical. They also found Gearheart's argument about the importance of nonpoint source pollution hard to swallow. Time would prove him right, but in the 1970s, water quality regulations were so focused on sewage treatment plants that even the most attentive bureaucrats had heard little about nonpoint pollution.

Had the hearing closed at the end of the first day, it's likely that the state board would have found in favor of the HBWA project. On the second day, however, the hearing room at the Elks Lodge was packed. Bertain had organized speakers from every corner of Humboldt County society: the Audubon Society and Sierra Club, timber companies, chambers of commerce, the bar pilots, the League of Women Voters. The range and depth of the opposition stunned Don Maughan and the other state board members as they sat through hours of irate testimony.[8]

Near the end of the day, a mysterious man from McKinleyville rose to testify; none of the longtime anti-HBWA activists recognized him. "You can't do this to us," he told the board. "Even if you convince all the rest, I'm going to sue you myself."

"The state board had been told that the opposition was just a vocal minority," recalled Ihara. "When they got here, it was everybody. Not just environmentalists, but exporters, average citizens, businesspeople."

Three weeks later, on May 17, Assemblyman Bosco announced the state water bureaucracy's surrender. In the aftermath of the Elks Lodge hearing, state board staff had written up a new report. Bosco read the juiciest quotes over the phone to a gob-smacked reporter for the Eureka *Times-Standard*. The report allowed that with proper treatment, sewage could be discharged to Humboldt Bay without causing bacterial pollution. "It is further concluded that projects such as the proposed Arcata marsh treatment process may enhance Humboldt Bay waters, as required by the Bays and Estuaries policy," it read. For Hauser, Gearheart, Allen, and Klopp, that single phrase represented victory after years of struggle.

The staff report insisted that the proposed HBWA project was a cost-effective solution to Humboldt's pollution problems, but acknowledged it would never be built. "Due to the widespread controversy and local opposition to the proposed regional project its timely and full implementation appears unlikely," it read. "Therefore, the board will consider other proposals, including those involving bay discharge, provided such alternatives provide for full secondary treatment and create new beneficial uses."

The same day, word came that the regional board would lift the cease-and-desist orders that had halted construction projects in Arcata and McKinleyville. Bosco withdrew the bill he had introduced, which would have rewritten the Bays and Estuaries policy. He had never intended to push it through the state legislature; it was meant as a crowbar that would pry the state away from its insistence on the HBWA project, and once that was accomplished, it could be put away.

Arcata was now free to act on its dream of wetlands treatment. The process would be shaped and sometimes blocked by state regulators, who had worked for years on the HBWA project and resented its fall.

NOTES

1. Hauser, D. (September 8, 1977). "Arcata vs. wastewater authority." *Arcata Union.*
2. Alm, A. (September 21,1977). "HBWA conflict still unresolved." *The Lumberjack.*
3. Anonymous (April 8, 1977). "Firm's chief claims HBWA ignored advice planning piping sewage under bay." *Times Standard.*
4. Fairbanks, A. (February 10, 1978). "Arcata wastewater plan is rejected." *Times Standard.*
5. Taylor, B. (February 15, 1979). "Moratorium hits—sewer battle keeps escalating." *Arcata Union.*
6. Taylor, B. (March 1, 1979). "HBWA battle on: initiatives are out, legislators are in." *Arcata Union.*
7. Taylor, B. (March 22, 1979). "Wastewater scuffles continue during week." *Arcata Union.*
8. Bretnall, P. (1984). "Wastewater conflict on Humboldt Bay." *Humboldt Journal of Social Relations* **11**(2): 128–284.

9

Do-It-Yourself Wetlands

Bob Gearheart emerged as Arcata's marsh guru during the city's long battle with the state water bureaucracy. This unpaid post demanded that Gearheart crank out proposals for wetland treatment at a frenetic pace, knowing that the city's financial future depended on his work. He wore a smile, energized by the pressure.

Gearheart's son, Greg, grew up to become an environmental engineer working for the state water board. He earned his engineering degree at Humboldt State, studying with his father. He remembers his dad happily engaged during the battle for Arcata's alternative treatment system, at the same time he was teaching a full load of classes. "My dad likes a fight," Greg says. "He adapts well. People put an obstacle in front of him, and he figures out a way to make it look like it's not really a problem. He makes it look like it was stupid on his opponent's part to put the obstacle there."

In 1977, the elder Gearheart proposed a first: a wetland built to treat municipal wastewater to the standards required under the Clean Water Act. He possessed a serene certainty that he could make this untried system work. "I had no data until we did the pilot study," he remembers, "but I was one hundred percent confident."

The power of aquatic plants to cleanse polluted water had first been tested in the 1950s by Käthe Seidel, a researcher at the Max Planck Institute in Germany. She showed that while some wild plants were killed off by waters tainted with phenol—a toxic organic compound used in making plastics—others had a remarkable ability to adapt. At first contact, effluent containing phenol caused bulrush stems to wither away, but the roots survived and in time sent up healthy new shoots. Bulrush, it turned out, could break down phenol, metabolizing it into the amino acids that build protein.[1] The plant also thrived in domestic sewage.

Seidel used carefully groomed cultures of wetland plants, rooted in beds of gravel or sand through which effluent flowed. Oxygen was pumped into the system, and because the sewage was not exposed to light, algae did not grow. She proved that in addition to breaking down toxic industrial pollutants, these reed beds could dramatically decrease the level of biological oxygen demand (BOD), suspended solids, and fecal bacteria in the water.

Based on Seidel's work, reed beds were successfully used to treat sewage from a Dutch campground. The bed killed off coliform bacteria and absorbed BOD, nitrogen, and phosphorus. The key to the system's function was the amount of time sewage spent filtering through the plants, known to engineers as the detention time. Sewage held in the reed bed for ten days or more was cleaner than the effluent from most conventional sewage treatment plants.[2]

At the time of Humboldt County's sewage rebellion, a handful of US researchers were studying the interaction of sewage and natural wetlands. One example was Philadelphia's Tinicum Marsh, on Darby Creek, near Delaware Bay. Before European settlement, six thousand acres of tidal marsh hosted abundant wildlife at Tinicum. Destruction of the wetland began in the 1630s, when Dutch, Swedish, and English colonists diked and drained the land for grazing. As the city grew, the wetland dwindled. In 1968, when construction of a new highway threatened to destroy the surviving 523 acres of marsh, conservationists set out to prove its value.

Three sewage treatment plants discharged into Tinicum, carrying an overdose of nutrients. The marsh was dominated by blooms of cyanobacteria and algae and a few pollution-resistant creatures, including crayfish and topminnows.[3] Still, an analysis of water quality showed that passage through the marsh reduced loads of phosphorus and nitrogen and raised oxygen concentrations. These findings helped in the fight to save the marsh, which was protected in 1970 as the first National Urban Park in the US.[4]

Several North American cities had been releasing treated sewage effluent into natural wetlands for decades—because it was convenient, not because wetlands were understood to improve water quality.[5] In the mid-1970s, researchers began to track the water quality impacts of wetlands, and documented the same pattern seen at Tinicum. When sewage passed through a swamp or marsh, concentrations of nitrogen, phosphorus, and coliform bacteria dropped, while dissolved oxygen rose.[6]

The pioneering wetlands ecologist Howard Odum had built a series of estuarine ponds near the sewage outflow from Morehead City, North Carolina, to observe the consequent changes in pond communities. Marsh grass and algae flourished in the ponds receiving wastewater, producing much more biomass than control ponds. Passage through the ponds filtered pollutants out of the sewage, improving water quality. Odum felt that wetlands could serve as a low-cost form of advanced sewage treatment.[7] He and his students demonstrated the concept by running sewage from a Florida trailer park through a natural cypress swamp. They found that the process worked well and that the swamp ecosystem was better able to handle high nutrient loads than open waters.[8]

John Teal at the Woods Hole Institution on Cape Cod studied the effects of sewage sludge experimentally applied to a Massachusetts saltmarsh. Though the sludge contained toxic metals, it carried a pulse of nutrients that caused plant growth to boom. A few species of pollution-resistant plants and animals flourished, and the marsh filtered out the majority of the contaminants.[9] One quote from Teal's work particularly struck Gearheart, and he would use it in the uphill battle to convince regulators that a manmade marsh could cleanse Arcata's effluent:

> Wetlands seem to be better processors of wastes than estuaries and coastal waters . . . [absorption of] nutrients, heavy metals, hydrocarbons, and

> pathogens are features of wetlands as they function naturally. They are in fact providing free waste treatment for contaminated waters already. It seems imperative therefore to implement wetland conservation to maintain this subsidy intact.[10]

The only existing lab data on manmade surface wetlands—as opposed to the subsurface flow reed beds used by Seidel—came from a study by National Aeronautics and Space Administration (NASA) engineers looking for a low-cost way to treat wastewater from the National Space Technology Laboratories in Bay St. Louis, Mississippi. In greenhouse studies, they found that dense stands of water hyacinth or alligator weed could remove more than 90 percent of the nitrogen and BOD in effluent, and more than 50 percent of phosphorus.[11] The NASA team envisioned constructed wetlands as part of a low-tech system that would sequester toxic metals and convert excess nutrients into a crop of water hyacinth that could be used as food. Later, wetlands would be proposed as part of a resource recycling system for manned space flight.[12]

Thanks to Dan Hauser's political savvy and funding from the Coastal Conservancy, Arcata's imagined marshes were becoming real. The city intended to integrate the marshes into its sewage treatment system, but the priority was to get them built as soon as possible. Hauser's pitch to the Conservancy, made while the HBWA battle was still in progress, underscored the urgent need for wetland habitat on Humboldt Bay. Since the region was settled during the California gold rush, more than 90 percent of native wetlands had been lost to diking and drainage.

Bulldozers began to dig basins for the marshes in September 1979, less than a year after Hauser had first applied for a grant. The groundbreaking was a full-on Arcata party, attended by the city council and staff, as well as an excited contingent from Humboldt State. A guy dressed as a tufted puffin waddled around in dive fins and sipped beer through an elaborate, thick-beaked mask. A buffet table was set up in the tall grass at the edge of Mt. Trashmore, along with a half-keg of local brew. "This is the culmination of a dream," Hauser announced as the bulldozers pushed into the earth.

Gearheart told reporters the marshes would serve as a national model for innovative sewage treatment. The full-scale wetlands had not been approved by state water quality officials, however, and for a few years would be fed with water pumped out of a nearby creek rather than treated sewage.

The marshes quickly proved their worth as wildlife habitat. Within a year, the first flooded marsh hosted American bittern, northern harrier, and an abundance of ducks. In the first spring, about two hundred ducklings—teal and mallards—hatched and fledged on the marsh.

Meanwhile, Gearheart had designed a pilot project to satisfy the state board's requirements. It would test effluent treatment in a dozen narrow marsh cells, using different plant species and flow rates. Soon after the celebration at Mt. Trashmore, a dozen trenches were dug in a diked-off corner inside the city's sewage plant. The cells were planted with different varieties of bulrush, and watered with effluent

Figure 9.1 Bob Gearheart at his namesake marsh in Arcata, 1989. Photo courtesy Campus Faculty Files, Humboldt State University Library.

from the oxidation ponds. Studies at the pilot marshes would carry on for more than twenty years, long after the marsh treatment system won approval (Fig. 9.1).

The lab at the pilot project was a decrepit single-wide mobile home. Gearheart's students busted out the interior walls and installed lab benches. The project's budget was minimal, and the main requirements for graduate students who worked on it were a sense of humor, a willingness to keep odd hours, and an ability to scrounge.

David Hull arrived as a fisheries graduate student in the summer of 1980. Gearheart hired him to run bioassays, a standard test in which fingerling salmon are put into sewage effluent for several days to test its toxicity. Hull needed dozens of glass tanks to do the work. "I had this brilliant idea," he remembers. "The Sparkletts water company was phasing out their five-gallon glass jugs, using plastic instead." He got the company to donate fifty of the glass containers. Using a homemade system of nichrome wire connected to a rheostat, Hull applied a jolt of electricity to cut the tops of the jugs off, like jack-o'-lantern lids. He had all the bioassay tanks he needed, free of cost.

In one corner of the lab, a computer geek named Nancy crunched water quality data on a Radio Shack computer with an eight-inch floppy disk, while in another, microbiologist Steve Wilbur peered at the tiny creatures that did the work of breaking down pollutants in the marsh. Wilbur was also a carpenter, who designed and built a greenhouse next to the trailer. The researchers used it to dry out plant samples, part of the process of measuring the biomass produced by the marsh cells.

The grad student who had spent the most time on the project was a wiry redhead named John Williams. He had been working with Gearheart since the

HBWA fight. For two years, Williams slept in the trailer cum laboratory to avoid the hassle and expense of paying rent on a place of his own. Gearheart's young kids called him Zonker because he looked so much like the Doonesbury character. His more official nickname was Marsh Man: He carried business cards inscribed with that title.

At one point the grad students decided to run a dye study, a way of tracking the speed at which wastewater moved through the marsh cells. They planned to take turns sampling each marsh cell hourly for 24 hours, but the water was moving slower than expected, and the process stretched out over five days of around-the-clock work. "You'd be out there in the middle of the night, falling in the test marshes because it was dark, and just having a good old time," recalls Hull.

"We got punchy," says Williams. "But we got it done."

Ask Hull about that time, and a wide grin spreads across his face. "We were a team, all in our twenties, making this project happen, working at the trailer at all times of the day and night." Over the months, they all learned about every aspect of the pilot and became adept at doing each other's jobs.

Bob Gearheart's son Greg was thirteen years old when the pilot marshes began running. Greg and his sister, Laura, spent a lot of time at the pilot project. "That trailer was like a clubhouse," remembers Greg. "There was a lot of work going on, and also a lot of fun." On the way to the marsh, the Gearhearts would stop at the Arcata Burger Bar. The woman who flipped the burgers there saved her gallon mayonnaise jars for the marsh researchers, who used them to collect samples. "I'm always happy to visit a treatment plant," Greg says, "because those places are so intertwined with my memories of eating delicious burgers and hanging with my dad."

By 1983, the pilot had demonstrated an ability to treat the city's effluent to a standard that met or improved on state water quality requirements.[13] The marsh's profound effect on water quality could be seen in samples of water taken at different points in the treatment process. Water that flowed out of the oxidation pond pea-soup green with algae emerged from the series of pilot marshes as clear as tap water.

"A marsh is a trickling filter set on its side," explains Williams. In a trickling filter, sewage is sprayed over a deep pile of rocks. A community of bacteria and fungi inhabit a biological slime coating the rocks, and do the work of breaking down organic matter. In a marsh, a similar slime coats the stems and roots of aquatic plants. Cattail and bulrush also pump oxygen into the system, helping the microbial community to thrive.

Studies on the pilot marshes increased Gearheart's understanding of how treatment wetlands work, and led to a major change in the design of the city's system. Instead of relying only on the three large enhancement marshes—dubbed Allen, Gearheart, and Hauser after the major players in Arcata's sewage rebellion—a series of densely vegetated treatment marshes were built beside the oxidation ponds. With their thick growth of cattail and bulrush shading the water, these marsh cells killed off algae, dramatically lowering suspended solids levels before the effluent was sent on to the larger enhancement wetlands. Arcata's natural sewage treatment system wasn't designed in advance, Gearheart acknowledges: It evolved, shaped by both political demands and scientific experimentation.

The key to obtaining the best water quality from marsh treatment, Gearheart wrote, was to keep the sewage in the system long enough for any suspended gunk to settle to the bottom and for the microbes that thrived amid the roots and stems of the cattail to break down organic matter. In engineering lingo, effective wetland treatment required detention time: days or weeks during which the habitat's natural processes could do their work. A long detention time also killed off most fecal coliform bacteria.[14]

In summer, the system worked beautifully, often releasing treated effluent that was cleaner than required by state standards. During the heavy rains of winter, millions of gallons of stormwater would seep into Arcata's aging sewer pipes. The volume of sewage increased dramatically, forcing operators to move the water through the marshes quickly, or to bypass them entirely. In the rainy season, the city often violated the standards set in its discharge permit. During the first years of the wetland treatment system, however, this seemed like a minor problem; the fine for an occasional violation was affordable.

By 1983, the wildlife marshes had become a popular attraction for birders, hikers, and school groups (Fig. 9.2). The regional board was ready to grant Arcata permission to discharge sewage effluent through the marshes and into Humboldt Bay on a long-term basis, on the premise that the marsh project enhanced the bay

Figure 9.2 River otter at the Arcata Marsh. Photo by Leslie Scopes Anderson.

in terms of both ecology and environmental education. The Department of Fish and Game, however, was opposed. Arcata's natural system didn't fit into state and federal regulatory schemes. Fish and Game biologists saw Allen, Gearheart, and Hauser marshes as habitats that had to be protected from pollution, rather than as part of the sewage treatment process. As for the regional board's wastewater engineers, they still refused to acknowledge the water quality benefits of treatment wetlands—despite the fact that Arcata's pilot marshes could produce an effluent similar to that released by high-tech tertiary treatment systems.

During the time he built Arcata's pioneering treatment wetlands, Gearheart was also spending months at a time traveling in the developing world to help create low-cost, low-tech systems to supply safe drinking water. In Ghana, he taught locals about simple filtration systems that could protect them from guinea worm infection. Among the Maasai in Kenya, he helped pinpoint the best spots to dig crevices that would store water in the dry season. He worked in Indonesia, Thailand, Sierra Leone, Tanzania, and Swaziland.

Gearheart encountered a successful low-tech, locally controlled water supply system among the nomadic Pokot tribe in Kenya. Tribal craftsmen had a long tradition of working metal. After a missionary taught them to weld, they began manufacturing high-quality water pumps. Each man signed his pumps, so that people would know who to find if it ever broke down. This simple approach gave people access to groundwater that was safe and free of pathogens. To show their appreciation for his help in locating the best places to dig wells, the Pokot presented Gearheart with a metal-tipped spear.

Gearheart saw that many water systems brought in by USAID and other First World agencies were too large, complex, and expensive for local people to maintain. For him, this made a clear parallel with Arcata's long struggle against the state water quality bureaucracy. Low-tech, local, sustainable technology made as much sense at home as it did in Africa or Asia.

Meanwhile, Hull and his fellow grad students continued to explore the mysteries of marsh treatment. Though the crew had carefully sown different types of plants in each pilot marsh cell, within a few months all the cells were dominated by a cattail, along with a scattering of bulrush. They soon realized that trying to enforce their experimental design on the plant community was a hopeless mission. "More important than the density and type of plants," Gearheart wrote, "is the distribution of plants in the flow pattern." The best arrangement for water quality purposes was a 60 to 80 percent cover of aquatic plants, and it was important that a dense stand be created adjacent to the outflow, to make sure suspended solids levels stayed low.

The pilot marshes cut the numbers of fecal coliform bacteria down to acceptable levels, without chlorination of the effluent, but this failed to impress the regional board's engineers. They insisted that Arcata's effluent be disinfected by chlorination twice: once before it was released to the enhancement marshes, and again before it flowed from the marshes to the bay. Chlorination reliably kills most pathogens and is standard practice at many wastewater plants. It also creates chlorinated organic compounds, which are toxic and persist in the environment.

Gearheart would spend decades struggling to get the chlorine out of Arcata's treatment system.

In 1986, Arcata finally received permission to flood its enhancement marshes with treated effluent. The influx of nutrient-rich water made aquatic plants grow like gangbusters. Marsh pennywort, a fast-growing native plant, formed floating mats that began to cover much of the surface, leaving little open habitat for waterfowl. Hull, who had finished his graduate work and been hired to manage the city's wetland system, continued his creative scrounging. He found a boat that had been abandoned on city property, painted it lime green, and rigged a rake on the bow. Puttering slowly through the marshes, Hull used the rake to rip out great mats of pennywort, which were later composted.

The Ford Foundation's Innovations in Government program awarded Arcata a $100,000 prize for its treatment system, calling it "an imaginative response to some of society's toughest issues."[15] Reporters showed up from *Time* magazine and the *New York Times*. Hull gave them grand tours of the wetlands. "It may be wastewater to you," he told them, "but it's our bread and butter."

As word spread of Arcata's successful experiment, Hull began to get inquiries from all over the world, asking how to build a treatment wetland. He tacked a poster-sized map of the world to his office wall and stuck in colored pins to represent the location of every person who had written to ask him how to create their own version of Arcata's system. The pins were scattered across the globe: Europe, Asia, Africa, South America. Hull remembers receiving queries written in Russian and carrying them to a linguist at Humboldt State to get them translated.

In a year or two, letters started arriving describing the success of far-flung treatment wetlands. One of the first projects to start running was in an indigenous Maori community in northern New Zealand. Wetlands not only saved money and energy, but they also fit much better with Maori traditions than conventional, mechanized treatment systems.[16]

The concept of treatment wetlands spread quickly. Wetlands are seldom practical for treatment of heavy loads of sewage in urban areas because they take up too much land in places where real estate is at a premium. In smaller and more remote communities, however, the technique is invaluable. In large swathes of Africa, Asia, and Central and South America, people have no electrical grid, no sewage pipes, and no money. A conventional treatment system requires expensive items like concrete, steel, and First World experts. Even when aid agencies donated state-of-the-art treatment systems, they never functioned for long because local people could not afford to maintain them. "Steel rusts away fast in the tropics," notes Steve McHaney, an engineer and one-time student of Gearheart's who has worked on treatment wetlands around the globe. By contrast, a wetland can be created using local materials and labor and will, for the most part, maintain itself.[17]

In 1992, a database recorded 127 treatment wetlands in North America. A year later, Gearheart helped the US Environmental Protection Agency (EPA) compile case studies of seventeen treatment wetlands in the US, including examples in Michigan, Oregon, California, Florida, Nevada, Arizona, and Illinois.[18] All of these were surface flow wetlands, similar to the Arcata marsh, though some were much larger. Hundreds of projects were treating wastewater in Denmark, France, Germany, and the United Kingdom. The European systems were subsurface wetlands, in which sewage flows through a bed of gravel or sand holding a dense stand of aquatic plants. Each approach had its advantages. Surface flow wetlands are simple to construct and provide important wildlife habitat. In subsurface systems, higher loads of contaminants can be treated in a smaller space. The wastewater remains underground, avoiding the risk of creating mosquito habitat, or of human contact with pathogens.

US engineers began to build wetland systems to treat water tainted with industrial pollutants, including benzene, aniline, and acid mine waste. They also began to use wetlands to manage a manmade excess of the nutrients nitrogen and phosphorus.

A few years after the seat-of-the-pants process that he used to create the Arcata marsh, Gearheart designed wetlands that required meticulous forethought and testing. A prime example was the Apache Superfund site in Cochise County, Arizona. The Apache Powder Company had manufactured nitrogen-based chemicals and explosives at the site for more than seventy-five years.[19] The factory released nitrate-laden wastewater to the surrounding desert, where it percolated into groundwater tapped by local wells. As a result, the groundwater carried high levels of nitrate, $NO_{3,}$ which can cause "blue baby" syndrome, a potentially fatal illness. The nitrate reacts with hemoglobin, rendering infants unable to carry oxygen in their blood—a condition that can cause coma and death if it's not treated.[20]

Gearheart designed a series of wetlands that would transform nitrogen, moving it from the water into the air. In small-scale tests, he found that bacteria required plenty of carbon to fuel the process of denitrification, which transforms nitrate into N_2 gas (see Chapter 3). At the Apache site, a series of wetland basins were dug in 1997, then planted with fast-growing cattail. It took a few years for the marshes to build up a layer of carbon-rich detritus from decaying plants. Even then, engineers had to add phosphorus to boost cattail growth, and molasses to increase carbon levels. When the treatment system finally went into full-scale operation in 2005, it proved able to take groundwater tainted with 200 mg/L of nitrate down to 10 mg/L, the federal drinking water standard.[21]

A breakthrough moment for emerging wetlands technology came in 1996 when the first comprehensive textbook on treatment wetlands was published. The lead author, Robert Kadlec, was a chemical engineering professor at the University of Michigan. He packed the text with mathematical formulae that helped to predict the complex biochemical interactions among plants, microbes, and pollutants. The book laid out principles that applied to surface as well as subsurface wetlands, and engineers on both sides of the Atlantic began to use both kinds of systems.

"Treatment wetland technology is growing exponentially," says Scott Wallace, an environmental engineer who is the principal of the consulting firm Naturally Wallace and has focused on wetlands since 1989. Wallace collaborated with Kadlec on a second edition of the treatment wetlands textbook, published in 2009.

"Frankly, it's beyond anyone's ability to count the number of systems now in operation," he says. "The global total is in the tens of thousands."

The design of each treatment wetland is shaped by the kind of waste to be treated and the local environment. Often, notes Wallace, engineering decisions come down to a tradeoff between the cost of land and the impacts of climate. Surface flow wetlands, which nurture wildlife and act as magnets for hikers and birdwatchers, remain the most common kind of system in North America.

Subsurface wetlands have advantages in extreme climates. Cold slows down the microbe-mediated reactions that break down organic matter and cycle nitrogen. In Minnesota, where winter temperatures can plunge to 25 degrees below zero, Wallace opted for a subsurface wetland insulated with mulch. In Saudi Arabia, where water evaporates quickly into the hot, dry air, subsurface wetlands are the only option.

The understanding that wetlands store and purify water has fueled some creative efforts to protect natural wetlands. Much of the domestic and industrial wastewater from the city of Kampala, Uganda, flows through the Nakivubo wetland before it reaches Murchison Bay, an arm of Lake Victoria. The lake suffers harmful algal blooms fueled by an overload of nutrients. Rapid population growth and skyrocketing real estate values in Kampala have driven a binge of wetland drainage and conversion. The Nakivubo filters the city's wastewater before it is released to Murchison Bay, just a few kilometers from Kampala's drinking water intake.

In the 1990s, a team led by economist Lucy Emerton of the International Union for the Conservation of Nature (IUCN) analyzed the monetary value of Nakivubo wetland's ecosystem services: its ability to break down organic matter, retain nitrogen and phosphorus, and kill off pathogens. Concluding that replacing these natural wetland functions would cost up to $1.75 million, the study found that Kampala could not afford to let the habitat be further diminished.[22] Emerton warned that because the great value of the wetland did not translate into private profits, halting the conversion of habitat would take aggressive planning and enforcement efforts.

City officials chose not to perform an expensive upgrade to Kampala's sewage treatment system and to rely on the Nakivubo wetland to purify its tainted waters instead. Unfortunately, loss of wetland habitat has accelerated: A recent study found that 62 percent of the area once covered by wetland vegetation was converted to croplands or buildings from 2002 to 2014.[23] As the population continues to grow, older drained areas on the marsh's edge are used to build houses or factories. Impoverished farmers are forced to move farther into the wetland to find moist soil on which to raise yams and sugarcane.

Wetland protection policies are on the books in Uganda, but they're not being implemented on the ground. The surviving segments of the wetland are contaminated with high concentrations of nutrients, coliform bacteria, and heavy metals. Recent evidence suggests that much of the natural filtering capacity of the Nakivubo wetland has been overwhelmed.[24]

In other parts of the developing world, constructed wetlands for water purification have boomed even as the loss of natural wetlands continues. The South China

Environmental Protection Agency built a demonstration treatment wetland in 1990.[25] Since then, China has embraced the new technology, studying ways to adapt its use from the tropical heat of Shenzhen to the chill northern provinces. In many small towns and rural areas, no conventional wastewater treatment facilities exist and constructed wetlands are of critical importance. Creation of treatment wetlands has accelerated in the twenty-first century, and much of the growing scientific literature on the subject now comes from Chinese researchers.

Environmental engineers have long referred to treatment wetland ecosystems as a "black box." "We've got a huge, functioning mess called wetlands out there with all sorts of interesting things going on inside it," Kadlec told a reporter in 1998.[26] "But we don't have enough information about what goes on inside the system . . . to advance our knowledge, we need to understand the internal processes that lead to the observed performance."

Researchers have begun to peer inside the black box. In the process, they have uncovered new and promising abilities of treatment wetlands to counter problems nobody was aware of in the days when Gearheart created Arcata's marsh system. Molecular fingerprinting has made it possible to track the ecology of distinct bacterial species involved in breaking down organic matter and cycling nutrients, and to discern the fate of pathogenic microbes.[27] Wetlands can filter out 99 percent[28] of fecal coliform bacteria (used as indicators of the potential presence of waterborne pathogens) through a combination of processes: bacteria settle into the sediments, or attach to plant roots. There they die off of starvation, of exposure to light or low oxygen conditions, or are eaten by nematodes, rotifers, protozoa, and other minute predators.

Bacteria are capable of promiscuously swapping genes, both within and between species. Conventional activated sludge treatment plants are hotspots where genes for antibiotic resistance spread, ultimately releasing resistant bacteria into the environment. Evidence suggests that constructed wetlands do a better job of killing off antibiotic-resistant bacteria and preventing the resistance trait from being transmitted—though the reasons remain unknown.[29]

Antibiotics are not the only drugs that enter the flow of wastewater and impact aquatic life; many pharmaceuticals emerge intact from conventional treatment systems. Examples include pain relievers and anti-inflammatory drugs like ibuprofen and naproxen, and synthetic estrogens used in birth control pills. Levels of synthetic estrogens in lakes and rivers are high enough to feminize male fish of many species, causing them to produce eggs instead of sperm. This syndrome has been documented in fish worldwide, and is widespread among bass in US rivers scattered from the Savannah in Georgia to the Yukon in Alaska.[30] Synthetic estrogens and other endocrine disruptors pose a profound threat to wild fish populations.

The mix of aerobic and anaerobic microhabitats in treatment wetlands can combine to aid in the breakdown of a range of pharmaceuticals—called "contaminants of emerging concern" by the EPA, which does not set effluent limits for these compounds as it does for BOD or suspended solids. Some drugs, like caffeine and ibuprofen, break down better under aerobic conditions. Hormone pollutants, like synthetic estrogen, can be degraded in both aerobic and anaerobic environments. Wetlands remove these contaminants through an array of processes: microbial breakdown, uptake and metabolism by plants, photodegradation.[31] Long retention times in wetlands increase the likelihood that pharmaceuticals will be broken down.

In the 1990s, scientists discovered a strain of bacteria in sewage sludge that was capable of cycling nitrogen in a previously unknown way.[32] These microbes grow slowly—one of the reasons they'd escaped identification for so long. They thrive in constructed wetlands, where a single cell can remain in place, metabolizing nitrogen and reproducing, for months.

The chemical trick these bacteria perform is known as Anammox, for anaerobic ammonia oxidation. In the absence of free oxygen, the microbes transform ammonia into harmless N_2 gas, removing nitrogen from the water. It's a radical twist on the classic model of the nitrogen cycle, which requires multiple species of bacteria and a mix of aerobic and anaerobic habitats to push nitrogen out of an aquatic environment. Because there's no need to push air into the system, as is done in activated sludge treatment, Anammox bacteria can take nitrogen out of wastewater using much less energy than conventional treatment.[33] The microbes have recently been harnessed to treat nitrogen-tainted water in specially designed reactors. But long before microbiologists had identified them, they'd been at work in wetlands.

The microbes that inhabit constructed wetlands live attached to sediments and to the underwater surfaces of plants, forming a biofilm. Levels of dissolved oxygen vary across the biofilm, allowing Anammox bacteria to coexist with the microbes that perform the classical nitrification and denitrification reactions—and conditions in a wetland can be manipulated to favor that coexistence. Together, the two classes of bacteria can dramatically reduce the amount of nitrogen in treated effluent.[34]

That's an important discovery, because wetlands are now being built to filter out the excess of nitrogen that modern humanity has unleashed on the planet. Lakes and estuaries in the US, and around the world, are choking on manmade overdoses of nitrogen and phosphorus. The nutrient overload runs off farm fields, lawns, and city streets—places where every scrap of wetland has long been drained or paved over.

NOTES

1. Seidel, K. (1976). "Macrophytes and water purification," pp. 109–121. In "Biological control of water pollution," Joachim Tourbier and Robert Pierson, eds. *University of Pennsylvania Press.*
2. de Jong, J. (1976). "The purification of wastewater with the aid of rush or reed ponds," pp. 133–139. In "Biological control of water pollution," Joachim Tourbier and Robert Pierson, eds. *University of Pennsylvania Press.*
3. Grant, R., Ruth Patrick (1970). "Tinicum Marsh as a water purifier." In "Two studies of Tinicum Marsh, Delaware and Philadelphia Counties, Pennsylvania." *Conservation Foundation* (19).

4. McCormick, J. (1971). "Tinicum Marsh." *Bulletin of the Ecological Society of America* **52**(3): 8, 23.
5. Kadlec, R.H., Scott D. Wallace (2009). "Treatment wetlands, 2nd edition." *CRC Press, Taylor & Francis Group, Boca Raton/London/New York.*
6. Boyt, F.L., S.E. Bayley, J. Zoltek, Jr. (1977). "Removal of nutrients from treated municipal wastewater by wetland vegetation." *Journal of the Water Pollution Control Federation* **49**(5): 789–799.
7. Mitsch, W., John Day, Jr. (2004). "Thinking big with whole-ecosystem studies and ecosystem restoration—a legacy of H.T. Odum." *Ecological Modelling* **178**: 133–155.
8. Odum, H.T., K.C. Ewel, W.J. Mitsch, J.W. Ordway (1977). "Recycling treated sewage through cypress wetlands in Florida." In "Wastewater Renovation and Reuse," Frank M. D'Itri, editor. *Marcel Dekker, Inc.* (http://ufdcimages.uflib.ufl.edu/AA/00/00/40/21/00001/Odum,HT,K.Ewel,W.Mitsch,J.Ordway.1975.Recycling%20Treated%20Sewage%20Through%20Cypress%20Wetlands.pdf).
9. Valiela, I., S. Vince, J.M. Teal, et al. (1976). "Assimilation of sewage by wetlands," pp. 234–253. In "Estuarine Processes," Martin Wiley, editor, Volume 1. *Academic Press.*
10. Ibid.
11. Wolverton, B.C., R.M. Barlow, R.C. McDonald (1976). "Application of vascular aquatic plants for pollution removal, energy, and food production in a biological system," pp. 141–149 in "Biological Control of Water Pollution," Joachim Tourbier and Robert Pierson, editors. University of Pennsylvania Press.
12. Nelson, M., A. Alling, W.F. Dempster, M. van Thillo, John Allen (2003). "Advantages of using subsurface flow constructed wetlands for wastewater treatment in space applications: ground-based Mars base prototype." *Advances in Space Research* **31**(7): 1799–1804.
13. Gearheart, R., Steve Wilbur, David Hull, Brad Finney, Shannon Sundberg (1983). "Final Report: City of Arcata Marsh Pilot Project." **1**(TD525.A 73. F 56); Gearheart, R. A., Frank Klopp, George Allen (1989). "Constructed free surface wetlands to treat and receive wastewater: pilot project to full scale." In "Constructed wetlands for wastewater treatment: municipal, industrial and agricultural," Donald Hammer, editor. *Lewis Publishers.*
14. Gearheart, R.A. (1981). "Use of vascular plants for treatment and reclamation of oxidation pond effluent and nonpoint source pollution loads." *Presentation, National Conference on Environmental Engineering, July 8–10, 1981.* Humboldt County Collection, HSU library (TD 755.G43).
15. Schneider, K. (January 24, 1988). "A town in California where waste isn't wasted." *New York Times.*
16. Tanner, C., James Sukias (April 2002). "Status of wastewater treatment wetlands in New Zealand." *EcoEng Newsletter* (http://www.iees.ch/EcoEng021/downloads/EcoEng021_F024_Wetlds.pdf).
17. Kadlec, R. H., Scott D. Wallace (2009). "Treatment wetlands, second edition." *CRC Press, Taylor & Francis Group, Boca Raton London New York.*
18. US Environmental Protection Agency (1993). "Constructed wetlands for wastewater treatment and wildlife habitat: 17 case studies." *EPA832-R-93-005* (https://www.cbd.int/financial/pes/usa-peswetlands.pdf).
19. Nelson, E., Leo Leonhart, Eric Roudebush, Kathryn Zaleski, Robert Gearheart (1999). "The Apache Wetland: design and construction," pp. 265–272. *Batelle*

Memorial Institute et al., International Wetlands and Remediation Conference, Salt Lake City, Utah.

20. Knobeloch, L., B. Salna, A. Hogan, J. Postle, H. Anderson (2000). "Blue babies and nitrate-contaminated well water." *Environmental Health Perspectives* **7**: 675–678.
21. Roudebush, E. M., Pamela J. Beilke (January/February 2006). "The Apache Nitrogen Wetland: groundwater denitrification using constructed wetlands." *Southwest Hydrology*, pp. 22–23.
22. IUCN (2003). "Nakivubo Swamp, Uganda: managing natural wetlands for their ecosystem services." *Case Studies in Wetland Valuation #7* (cmsdata.iucn.org).
23. Isunju, J. B., Jaco Kemp (2016). "Spatiotemporal analysis of encroachment on wetlands: a case of Nakivubo wetland in Kampala, Uganda." *Environmental Monitoring and Assessment* **188**: 203.
24. Fuhrimann, S., Michelle Stalder, Mirko Winkler, et al. (2015). "Microbial and chemical contamination of water, sediment and soil in the Nakivubo wetland area in Kampala, Uganda." *Environmental Monitoring and Assessment* **187**: 475.
25. Du, C., Zongyan Duan, Baokun Lei, Wanli Hu, Shihua Chem, Guimei Jin (2014). "Research progress on application of constructed wetland in wastewater treatment in China." *Agricultural Science and Technology* **15**(2): 310–320.
26. Cole, S. (May 1, 1998). "The emergence of treatment wetlands." *Enviromental Science & Technology*, pp. 218A–223A.
27. Vacca, G., Helmut Wand, Marcell Nikolausz, Peter Kushk, Matthias Kastner (2005). "Effect of plants and filter materials on bacteria removal in pilot scale constructed wetlands." *Water Research* **39**: 1361–1373.
28. Ghermandi, A., D. Bixio, P. Traverso, I. Cersosimo, C. Thoeye (2007). "The removal of pathogens in surface flow constructed wetlands and its implications for water reuse." *Water Science and Technology* **56**(3): 207–216.
29. Wu, S., Pedro Carvalho, Jochen Muller, Valsa Remony Manoj, Renjie Dong (2016). "Sanitation in constructed wetlands: a review on the removal of human pathogens and fecal indicators." *Science of the Total Environment* **541**: 8–22.
30. Hinck, J., V.S. Blazer, C.J. Schmitt, D.M. Papoulias, D.E. Tillitt (2009). "Widespread occurrence of intersex in black basses from US rivers, 1995–2004." *Aquatic Toxicology* **94**(4): 60–70.
31. Verlicchi, P., Elena Zambello (2014). "How efficient are constructed wetlands in removing pharmaceuticals from untreated and treated urban wastewaters? A review." *Science of the Total Environment* **470**: 1281–1306.
32. Kuenen, J.G. (April 2008). "Anammox bacteria: from discovery to application." *Nature Reviews Microbiology* **6**: 321.
33. Wallace, S., David Austin (2008). "Emerging models for nitrogen removal in treatment wetlands." *Journal of Environmental Health* **71**(4): 10–16.
34. Tao, W., Jianfeng Wen, Youl Han, Matthew Huchzermeier (2012). "Nitrogen removal in constructed wetlands using nitration/anammox and nitrification/denitrification: effects of influent nitrogen concentration." *Water Environment Research* **84**(12): 2099–2105.

10

Strangled Waters

Second Wave

In August 2014, the water supply for the city of Toledo, Ohio, was poisoned. Officials issued an order to the half-million residents connected to the municipal water supply: Don't drink, cook, or brush your teeth with the water. Do not use it to bathe your children, and don't give it to your pets.[1] Stores ran out of bottled water, and residents had to wait in long lines or travel to neighboring towns to find more.

The culprit was a bright green plume of *Microcystis*, a cyanobacterium that thrives in warm water tainted with heavy loads of phosphorus and nitrogen.[2] Every spring, rains wash a pulse of nutrients off fertilized fields and send it down the Maumee and Sandusky rivers and into western Lake Erie. Every summer, as water temperatures rise, *Microcystis* forms an iridescent mat over parts of the lake's surface.[3] In early August 2014, strong winds blew a lawn of cyanobacteria over Toledo's water intake, which lies just outside the Maumee's mouth. Tests showed that the city's water contained dangerous levels of microcystin, a liver toxin produced by the bloom. The drinking water crisis was a dramatic signal of Lake Erie's descent back into eutrophication.

In the 1980s, after sewage plants in the watershed were upgraded and phosphate detergents banned, Lake Erie experienced a revival. Algal blooms faded, and populations of walleye rebounded. The lake grew a thriving tourist industry based on sport fishing. Then, in 1995, researchers recorded the lake's first widespread bloom of *Microcystis*.

Eruptions of *Microcystis* have since become a predictable event striking the western Lake Erie basin every summer. The most widespread and long-lasting blooms hit in 2011 and 2015, after intense spring rains dumped heavy loads of nutrients into the lake. Climate models forecast warmer summer temperatures and heavier spring rains for the Great Lakes region. Those conditions are a recipe for more and larger algal blooms, and are likely to favor *Microcystis* in particular.[4]

The regulatory efforts of the 1970s and 1980s made great progress in cleaning up discharges from industries and sewage treatment plants, but failed to address nonpoint source pollution flowing from farm fields and city streets. An overload of nutrients carried in storm runoff from cropland is now the major source of pollution in Lake Erie's western basin.

An emerging body of research suggests that human actions have driven the evolution of new and different kinds of harmful algal blooms (HABs)—or more specifically cyanobacterial blooms (cyanoHABs). *Microcystis* was present in the twentieth-century lake, during the days when Lake Erie was declared "dead" by the popular press, but back then it was relatively scarce. Today it dominates cyanoHABs not only in Lake Erie but in many polluted waters worldwide, including Florida's Lake Okeechobee, the Baltic Sea, Lake Taihu in China, and Lake Ohnuma in Japan. *Microcystis* blooms have been reported from 257 countries in the Americas, Europe, Asia, Australia, and Africa.[5] The organism thrives in warm waters that are rich in nutrients and low in salinity.

Delving into the ecology of *Microcystis*, scientists have begun to debunk the long-held assumption that control of phosphorus alone can rescue eutrophic lakes. Phosphorus loads do fuel cyanoHABs in Lake Erie, but excess nitrogen shifts the balance in favor of *Microcystis*, rather than diatoms, green algae, or the nitrogen-fixing cyanobacteria that dominated polluted lakes in the twentieth century. Microcystis also requires abundant nitrogen to produce toxin. "Many lakes that have *Microcystis* blooms are receiving increasing loads of nitrogen from synthetic fertilizers, urban runoff, and atmospheric pollution," says Hans Paerl, a microbial ecologist at the University of North Carolina at Chapel Hill. "Nitrogen is the new part of the story."

The principal source of both phosphorus and nitrogen flowing to Lake Erie stretches for thousands of square miles across northwestern Ohio and eastern Indiana. The rich earth and the hard work of farmers in the region produces hundreds of millions of dollars' worth of soybeans and corn, as well as wheat, vegetables, pork, and poultry. The landscape is a vast, flat expanse of tidy fields dotted with modest farmhouses and criss-crossed with county roads.

The Great Black Swamp once stood here. A missionary who passed through in 1791 described "deep and troublesome marshes, the horses at every step wading up to their knees."[6] On an 1808 map, the swamp, which covered most of northwestern Ohio, was designated as "land not worth a farthing." Settlers came anyway, felling the giant sycamores and oaks, strapping wide wooden shoes to their horses' feet to stop them from sinking into the mud. By the turn of the twentieth century, blood, sweat, and ingenuity had transformed the swamp into productive farmland.

The tough people who conquered the Great Black Swamp passed down a loathing of wetlands that remains strong to this day. "Ohio has eliminated a large percentage of our natural wetlands," says Sandusky County engineer Jim Moyer. "Well, I say *good*, because that's where I live." These attitudes are written into state law, which makes impossible any action, including wetland creation, that slows the flow of runoff through constructed drainage ditches—the conduits that, after each heavy rainfall, deliver thousands of metric tons of phosphorus and nitrogen[7] to the Maumee, and onward to Lake Erie.

In the aftermath of Toledo's drinking water crisis, wetlands expert and professor emeritus at Ohio State University William Mitsch offered up a characteristically blunt manifesto. Harmful algal blooms have become chronic in western Lake Erie, he wrote, because of the wrongheaded way people manage the watershed.[8] Mitsch proposed that a significant fraction of the region's rich cropland be taken out of production and converted to wetlands, which act as natural filters for the overload of nutrients now flowing from farm to river to lake.

Mitsch's idea strikes many locals as pure craziness, but that doesn't bother him. A bearded, bespectacled man who knows wetlands in the intimate way most of us know our hometowns, he built a wetland park at the Ohio State campus in Columbus, where he spent more than twenty years teaching and studying ecological engineering. He coauthored a textbook that has become a bible for students of wetland ecology and management—now in its fifth edition.[9]

With an encyclopedic knowledge of close to fifty years of evolving science, Mitsch has become an eminent advocate for wetland restoration in a world where these habitats have become rare and precious (Fig. 10.1). His dream for the former Great Black Swamp is based on a lifetime of studying the inner workings of wetlands.

He envisions tens of thousands of acres—or roughly 7 to 10 percent of the ground here—being restored to wetland. "You'd need 10 percent of the landscape filtering for you so you can farm the other 90 percent," Mitsch says.

That's going to be a tough sell to the region's thousands of farmers, who make their living off the soil laid down by the former Black Swamp. For them, says

Figure 10.1 Bill Mitsch at the Everglades Wetland Research Park, Naples, Florida. Photo by Sharon Levy.

Kris Swartz, a farmer and president of the Ohio Federation of Soil and Water Conservation Districts, every bit of earth is valuable. Swartz estimates that it would take a forty-acre wetland for every square mile of crops to solve the nutrient pollution problem in western Lake Erie.

"Who's going to give up the ground?" he asks.

As a graduate student in the 1970s, Mitsch ran sewage from a Florida trailer park through a native cypress swamp.[10] His experiment demonstrated that the wetland acted as an efficient natural filter, absorbing pathogenic bacteria, organic matter, nitrogen, and phosphorus. At the time, researchers in Michigan were showing that peat bogs could serve the same function. Mitsch's blue-gray eyes gleam behind his wire-rimmed glasses as he remembers that era. "We discovered this ecosystem service of wetlands cleaning polluted water," he says, "and interest in them exploded. People began to argue that wetlands should be protected because they act as nature's kidneys."

Mitsch's textbook includes a map of the original Black Swamp. He sees it as an important example of a landscape transformed, to the point where few remember it was ever a wetland. It's a pattern that has played out more than once in American history: The nation's most productive farming areas are converted wetlands. Much of the Midwestern Corn Belt was wet prairie. California's Central Valley was a vast tule marsh. The sugar fields of south Florida were a mix of cypress swamp and sawgrass wetland.

After settlers cleared the Black Swamp of its giant trees and drained it, they found themselves in possession of some of the most productive farmland on the planet. Over time, farming depleted the soil of nitrogen and phosphorus, elements key to plant growth. So farmers added manure, and later, synthetic fertilizer. Some of the nutrients were taken up by crops. Much washed down to the ditches, into the Maumee and Sandusky rivers, and on to Lake Erie.

Today, farmers in the former swamp continue to lay more lines of drain tile, spaced closer together. The clay pipes placed by their ancestors have broken; the modern version consists of plastic tubing that can be purchased by the spool.

Rainwater that percolates through the soil of farm fields and into the tile drains dissolves phosphorus. Conventional wisdom once held that phosphorus ran off agricultural land only in particulate form, attached to grains of sediment. Particulate phosphorus tends to settle out of the water column and remain, for the most part, out of reach for nutrient-hungry algae. Soluble phosphorus, known in the lingo of water pollution as dissolved reactive phosphorus (DRP), is 100 percent bioavailable.

In the mid-1990s, levels of DRP began to climb in the tributaries of the Maumee.[11] DRP fuels blooms of *Microcystis* and other kinds of cyanobacteria that have become an annual blight in Lake Erie.

The rise in DRP was triggered by a series of changes in farm management.[12] Instead of rotating their fields among a number of crops, including hay, wheat, and rye, farmers focused on highly fertilizer-dependent corn and soybeans. Subsidies for ethanol production drove yet more planting of corn. Many northwest Ohio farmers adopted no-till agriculture, in which seeds are planted through the stubble of the previous year's crop, without plowing. No-till decreases erosion and improves the health of the microbial communities that build soil. Soil in no-till fields is coarse, and tends to slow the flow of water, allowing it to seep underground and decreasing surface runoff. At the same time, it increases the amounts of dissolved nitrogen and phosphorus reaching tile drains.

In 1995, the first major bloom of *Microcystis* struck thewestern basin,[13] fueled by nutrient-rich farm runoff flowing out of the Maumee. The river that once ran clear through the northern reaches of the Great Black Swamp now carries an overdose of phosphorus and nitrogen to the lake every spring, triggering *Microcystis* blooms every summer.

"Replace a swamp with high-fertilization agriculture, and you're going to screw up something downstream," says Mitsch.

Lake Erie's most intense bloom to date occurred in the summer of 2015, after a spring of heavy rain. Climate change models predict heavier spring rains and higher summer temperatures for the region, a recipe for intensified blooms of cyanobacteria.[14] But even if reviving parts of the Great Black Swamp makes sound ecological sense, it's an unpopular idea in an area devoted to farming.

"Mitsch is right in concept," says Ken Krieger, director emeritus of the National Center for Water Quality Research at Heidelberg University in Tiffin, Ohio. "But it's very unlikely that politically and socially the climate will ever be right to take land away from farming and convert it back to wetlands."

The most impressive remnant of the Black Swamp is Goll Woods Nature Preserve. An island of forest that juts from a sea of farm fields in Fulton County, Ohio, the preserve holds some of the last old-growth trees that grew in the original swamp. Centuries-old bur oaks and sycamores stand like the abandoned pillars of a long-vanquished kingdom. Their massive trunks run straight toward the sky, supporting whorls of leafy branches that hang ninety feet up in the air. Standing in the heart of Goll Woods, immersed in the aroma of dry mud and fallen leaves, it's easy to conjure the lost wilderness. An expanse of ancient trees stretched for a million wet acres, interrupted only by stands of tall marsh grass. The residents were deer, elk, wolf, cougar, badger, beaver, bear, muskrat, and an astounding abundance of waterfowl. But in northwest Ohio, the swampy past is a very foreign country.

A federal program devoted to controlling nutrient pollution in Lake Erie's western basin supports small wetland restoration projects. The program allows the government to rent fields that are converted to marsh, and to pay landowners rates that can make at least marginal, flood-prone land more profitable than struggling to grow a crop. But relatively few farmers have signed up, and those who do hear complaints from neighbors who fear any wetland will cause water to back up and flood their crops, or who simply can't abide the untamed look of wetland plants.

"We're putting wetlands in," says Steve Davis, a watershed specialist with the federal Natural Resources Conservation Service. "Not near as many as we would want to, or as we could, or even as we have funding for." Most of the landowners who accept the idea are duck hunters, motivated to provide habitat on their property.

Brothers Christian and Peter Lenhart co-own a farm near Defiance, Ohio that's been in the family since it was first homesteaded in the 1840s. When Chris, a water resources scientist at the University of Minnesota, first brought up the idea of restoring a marsh on part of the land, local agricultural officials tried to talk him out of it.

"Ohio is much more anti-wetland than the rest of the Midwest," notes Chris, who works with Minnesota farmers to develop wetlands designed to filter agricultural runoff. "Talk to people about building a wetland in Defiance County, and they kind of think you're crazy. It goes beyond rationality. It's the communal memory of people dying of malaria in the Black Swamp."

Pete, a physician in Defiance, has grown passionate about both native plants and wetlands. To create the seasonal marsh that now thrives on the Lenhart farm's back forty, he and his daughters hand-collected seeds of rare native plants. Walking there on a sunny July afternoon, Pete runs his hand lovingly over the bright yellow blossoms of black-eyed Susan, the curling purple petals of bergamot, the bunched seedheads of Ohio spiderwort. The sedges he planted flourish on the wetland's floor, which goes dry in summer.

In spring, this ground lies beneath a sheet of water and attracts an abundance of dabbling ducks—blue-winged teal, northern shovelers, pintails. During their long migration north to breeding grounds in Canada, the birds drop down to rest on the Lenharts' wetland, a rare oasis of natural water among the drained fields. The northern harrier, a marsh-adapted hawk now endangered in Ohio, has been seen hunting here. In spring the neighbors complain of the noise from hundreds of courting frogs.

The Lenhart wetland clearly works as habitat,[15] but in terms of improving the quality of water running down to the Maumee, it's not doing much. The family could not build a wetland that intercepted the flow of polluted runoff in the ditch, because the Defiance County Soil and Water Conservation District, which is responsible for enforcing state drainage laws, rejected the idea. "They're very authoritarian and incredibly powerful," says Pete, whose sardonic shorthand for drainage officials is "ditch Nazi."

The district recently came in and rehabbed the ditch that passes through the farm, over the objections of the Lenharts and other local property owners. Before the district improvements, Pete could count around fifty species of plants growing in the ditch—many of them natives. Now the area is dominated by a single species: fescue.

Jason Roehrig, administrator for the Defiance County Soil and Water Conservation District, points out that drainage ditches are manmade channels designed to keep water off the fields. By law, ditches under the district's purview must be maintained in perpetuity—not as habitat for native plants but as stable conduits that can handle intense flow in times of heavy rain. Fescue is not a native species, but Roehrig notes that it "provides a durable, dense sod that resists erosion and is relatively quick and easy to establish." By contrast, native plants are much slower to take hold.

"These fields aren't ecosystems," Pete says as he gestures at the vista of drained fields that stretches to the horizon. "They're factories." During my travels in northwest Ohio, I heard the same sentiment from several other farmers.

Ohio's ditch regulations give every landowner a right to the fastest, most efficient drainage, and make no allowance for other priorities. "The ditch commissions are in direct conflict with efforts to clean up the water flowing into Lake Erie," says Pete.

Jim Moyer, the engineer for Sandusky County, has spent much of his career designing and maintaining drainage ditches. One of the biggest challenges in his work is the flatness of the landscape. If the fields of the former Black Swamp were significantly uphill from Lake Erie, he notes, it would be easier to build wetlands into the ditch system without disrupting it. Moyer has boated through bright green algal blooms in Erie's western basin. "I don't want to swim in this lake if it's gonna be like that. I don't want to eat any fish out of it either," he says. "So I hope we come up with a solution."

He doesn't see large-scale wetland restoration as practical, however. "I try to be realistic," he says. "In our county, agriculture is important. We have corn, soybeans, and wheat, but also tomatoes, cucumbers, peppers, cabbage: those are high-dollar crops, and, boy, they cannot tolerate any excess water."

Most farmers in the region acknowledge the need to limit nutrient runoff. The federal Natural Resources Conservation Service has a three-year, $41 million program available to help farmers implement changes aimed at stopping nutrient pollution in western Lake Erie. The money can be used to fund a range of strategies, from high-tech management of fertilizer application to planting buffer strips of grass or trees at the edge of ditches. Buffer strips have been widely adopted in the former Black Swamp. They are effective in decreasing erosion, and can reduce the amount of total phosphorus running off a field by about half.[16] Tile drains running beneath the soil surface continue to carry soluble phosphorus and nitrogen off the land, however.

A recent analysis by researchers at the University of Michigan suggests that under the current levels of use, buffer strips and other best management practices (BMPs) aimed at minimizing nutrient runoff won't be enough to rescue Lake Erie.[17] The BMPs do help, but a relatively small proportion of farmers are using them.

The study found that taking 50 percent of cropland out of production and letting switchgrass grow there would bring phosphorus loads in Lake Erie down by 40 percent, the goal set in an international agreement between Canada and the US. The authors acknowledge that's never going to happen, but they included that example in their study to underscore the scale of change that will be needed to address the problem.

Mitsch agrees that BMPs alone won't solve the problem. He points to the infamous dead zone in the northern Gulf of Mexico, which is fueled by nutrients that run off farms in the Mississippi basin. In the 1990s, Mitsch was part of a committee that studied solutions to the problem, and concluded that BMPs were part of the answer, along with building wetlands back into the watershed. Twenty years later, only a few scattered wetlands have been restored. In 2017, the dead zone expanded to an unprecedented 8,776 square miles, about the size of New Jersey.[18]

Chris Lenhart has begun studying nutrient retention in the family's marsh and two other nearby wetland projects. "Our wetland is removing a fair amount of nitrogen and phosphorus," he says, "but it's not treating any additional farmland. To really treat the problem, you need wetlands that collect runoff from hundreds of thousands of acres."

There is only one place on the planet where constructed wetlands are successfully treating farm runoff on that scale: in the Everglades Agricultural Area (EAA) of south Florida. Once part of the sawgrass marsh and cypress swamp that made up the original Everglades, the EAA was drained in the early 1900s, when canals were dug to divert Lake Okeechobee's waters to Florida's coasts. As in the Black Swamp, drainage uncovered fertile wetland soils. Farm runoff degraded the surviving Everglades wetlands, which lie south of the EAA. Cattail, a wetland plant adapted to phosphorus-rich waters, displaced the native sawgrass. Herons and egrets find little prey in thick stands of cattail, and the plant's rapid expansion contributed to a 90 percent decline in the abundance of wading birds in the Everglades.[19]

By the 1980s, conservationists were demanding that Everglades National Park be protected from water pollution. In 1988, the federal government sued two Florida environmental agencies for failing to enforce the state's own water quality standards. Three years into the litigation, Governor Lawton Chiles walked into federal court in Miami and announced the state's surrender. "We want to plead that the water is dirty," he said. "We want the water to be clean, and the question is how can we get it the quickest."[20]

After more than two years of intense negotiations, the US Department of Justice and the state of Florida reached a settlement agreement. The path to clean water in the Everglades would involve converting tens of thousands of acres of farmland to constructed wetlands, designed and managed to filter out phosphorus. In addition, farmers would be required to use BMPs to limit the concentrations of phosphorus in runoff.

The first pilot treatment wetland was created on 3,815 acres of state-owned land. Scientists working on the project demonstrated that the managed marsh could remove about 80 percent of the incoming phosphorus load. Over time, cattail and other aquatic plants died and decomposed on the bottom, building a new layer of organic peat that acts as long-term phosphorus storage.[21]

The lessons of the first pilot marsh have been applied to a series of scaled-up constructed wetlands built at the southern edge of the EAA, called stormwater treatment areas (STAs) by the South Florida Water Management District. The STAs now cover more than 57,000 acres.

"The STAs do an outstanding job of filtering phosphorus," says Gary Goforth, an environmental engineer who worked on development and operation of the large-scale wetlands for two decades. When the first STAs were built in 1994, the goal was to take runoff water carrying about 200 parts per billion (ppb) phosphorus and reduce the load to 50 ppb. Nowadays, the STA effluents are well below that figure, often containing less than 20 ppb phosphorus. Walk one of the dikes that surround the wetland cells, and you'll find yourself surrounded with wood storks, roseate spoonbills, black-necked stilts and alligators.[22]

The success of the STAs has been possible in part because farmers adopted a range of BMPs, including holding water on some of their fallow fields and carefully measuring soil nutrient levels so that fertilizer is applied only where it's needed. The Everglades farming community had to be drawn into the process against its collective will. Goforth, who served as an expert witness during a long series of court cases involving Everglades restoration, recalls farmers testifying that a required 25 percent reduction in runoff phosphorus loads would put them out of business. Today, using an evolving array of BMPs, phosphorus loads in waters entering the STA system have been reduced significantly—by an average of about 50 percent in recent years.

The engineers who built the STAs and continue to monitor their function have created an extensive knowledge base on how to design and operate large treatment wetlands. Circumstances would, of course, be different in northwest Ohio, where wetlands might freeze and plant growth halts during winter. The system should be easily adapted, however.

Mitsch has been tracking two small restored wetlands he built on the Ohio State campus in 1994. They are reliably retaining phosphorus, carbon, and nitrogen. Robert Kadlec, an emeritus professor at the University of Michigan who has consulted on the design and operation of the STAs since their inception, has also demonstrated that both natural and constructed wetlands effectively absorb excess nutrients in the Midwest.[23] "What we need in the western Lake Erie watershed," says Mitsch, "is something on the scale of the STAs in Florida."

The major obstacles, whether in Ohio or Florida, are social and political. "In my experience," says Goforth, "agriculture is a very strong lobby in any state, any country. It took the federal lawsuit to bring the farmers to the table. They realized it wasn't just business as usual."

Farmers in the EAA have adapted. Keith Wedgeworth runs a farm that has been in his family since the 1920s. "We bought into the STAs, and they've done what they were supposed to," he says. Due to an array of BMPs, Wedgeworth says the runoff leaving his farm is cleaner than the water that flows in from Lake Okeechobee.

After Mitsch retired from teaching at Ohio State, he returned to south Florida. He is now director of the Everglades Wetland Research Park at Florida Gulf Coast University—living and working in America's wetland mecca. The struggle to create the STAs there was resolved when the state was able to buy land from willing sellers. Wetland peat makes rich soil, but when dried out and exposed to the air it can oxidize, vanishing in a slow, invisible burn. At the southern edge of the EAA, farmers were going out of business because the layer of soil above the limestone bedrock had worn too thin. They were willing to move on.

Still, the controversy has recently revived, because it turns out 57,000 acres of treatment wetlands may not be enough. Heavy rains struck Florida in the winter of 2015–2016, during what is usually the region's dry season. The volume of polluted water that ran into Lake Okeechobee was too much for the STAs to handle, so it was sent, untreated, down the canals that lead to the Atlantic and Gulf coasts. The stormwaters carried an intense pulse of nutrients to estuaries, triggering a guacamole-thick bloom of cyanobacteria that continued to plague the region months later.[24] Mitsch has suggested that the only solution is to buy out more than 80,000 additional acres of farmland to create more STAs.

If that is ever to happen, the question will be the same one Swartz poses in Ohio: Who will give up their ground? Wedgeworth feels that farmers in the EAA, south of Lake Okeechobee, are already doing their part. "We're part of the solution," he says. "It's in my interest to be the best steward of the land possible, so I'm able to pass it down to future generations in my family." The nutrient load tainting Lake Okeechobee flows in from the north, and Wedgeworth argues that any new water treatment projects should be built upstream.

Everglades National Park covers 1.5 million acres and is known worldwide as a wilderness gem. The Everglades is listed as a Wetland of International Importance under the Ramsar Convention, a pact aimed at protecting vital wetland habitats worldwide. The push for a global wetlands agreement began in the 1960s, fueled by concerns over vanishing habitat for migratory waterbirds, but Ramsar Convention criteria have evolved to recognize other wetland ecosystem services, including water quality protection.

Conservation groups in Ohio, meanwhile, have focused on restoring native wetland plants and providing waterfowl habitat. Altering flows to allow polluted water to slow down, spread out, and be filtered through aquatic vegetation is rarely considered, Mitsch says. The only Ramsar site in Ohio is the Olentangy River Wetland Research Park, which takes up fifty acres on the Ohio State University campus in Columbus. Mitsch drove the creation of the wetland complex beginning in 1994, when he built his two small, experimental wetlands fed with water pumped from the Olentangy.

Mitsch remains hard at work, trying to demonstrate that farming and wetlands can coexist. His latest study uses mesocosms, arrays of miniature marshes set up in tubs, which allow him to test the effects of different flow rates and plant communities on a wetland's ability to take up nutrients. One test site is near the shore of Buckeye Lake in Ohio, which has been plagued with harmful algal blooms. A second is in south Florida, and Mitsch is seeking funds for a third site on Long Island, New York.

"We're going to experiment with a clever way of trapping phosphorus in wetlands and then returning it to agriculture," says Mitsch. He won't reveal the details of the plan, which he says are proprietary—but it will be applicable to the former Black Swamp. The goal is to merge wetlands with farming, using manmade marshes to capture nutrients that can then be returned to the fields, saving farmers the cost of synthetic fertilizer.

Mitsch was encouraged when he visited Sweden and was shown a system set up to capture the nutrient pollution that has caused harmful algal blooms in the Baltic Sea. He sent me a photograph: a flat green field that might have been in northwest Ohio. What marked the place as foreign was a long strip of tall reeds, sprouting from a constructed wetland that stands between the field's edge and the drainage ditch. Swedish researchers have suggested that the government should pay farmers based on the amount of nutrients sequestered in their wetlands, a policy that would cast farmers as environmental entrepreneurs.[25]

Mitsch has been thinking big about wetlands for a lifetime. He still hopes to transform the way soggy American landscapes have been managed since settlement times, and to fit wetlands back into the picture. He insists that any real solution will have to be deployed on a grand scale, one that matches the sweeping changes made to the former Black Swamp. If we're not willing to do that, Mitsch says, we should admit we're willing to accept lakes painted bright green by blooms of cyanobacteria.

In estuaries, where rivers meet the sea, the joining of fresh and salt waters traditionally results in a profusion of life. But the same plague that has struck Lake Erie now plays out in estuaries around the globe: HABs create large areas of oxygen-depleted water in more than four hundred estuaries worldwide, where major watersheds empty their loads of human-generated nutrients into the sea.[26] In the years after World War II, the use of synthetic nitrogen fertilizers increased dramatically. The planet's coastal dead zones have expanded and intensified over the same stretch of time, fed by a human-generated glut of nitrogen.[27] (In salt water, nitrogen is most often the crucial limiting nutrient.)

A repeated pattern of destruction has been traced in eutrophic estuaries around the world. An overload of nutrients leads to blooms of phytoplankton, dominated by toxic forms that cannot be eaten by copepods and other small grazers. Meanwhile, over-harvest of larger fish, often the apex predators, means that smaller prey fish multiply, feeding on and further reducing populations of zooplankton. So phytoplankton flourish, forming larger HABs. Often an invasive marine animal tolerant of eutrophic conditions will come to dominate. When the Black Sea's ecosystem collapsed in the 1980s, native fish vanished and invasive comb jellyfish became wildly abundant.[28]

Remnants of long-dead foraminifera, tiny marine creatures that leave behind ornate shells of calcium carbonate, reveal the history of nutrient overload in great estuaries like the Chesapeake and the northern Gulf of Mexico. The diversity of foraminifera (forams for short) declined in the Gulf during the 1940s as the use of synthetic nitrogen fertilizers began in earnest. The ratio of *Ammonia parkinsoniana*, a foram tolerant of low oxygen conditions, to that of oxygen-hungry *Elphidium* spp. has proved to be a consistent sign of oxygen depletion in the Gulf, Chesapeake Bay, Long Island Sound, and other coastal areas.[29] In all these waters the ratio of *Ammonia* to *Elphidium* increased significantly after the 1950s. An oxygen-loving foram, *Quinqueloculina*, was abundant in the Gulf from 1700 to 1900, showing that hypoxia was not a problem then. Today the dominant foram is a hypoxia-tolerant creature, *Buliminella morgani*, known only from Gulf of Mexico.[30]

River flows form a lens of lower-density water as they empty into an estuary, trapping the saltier bottom waters below. As phytoplankton bloom and die, their

Figure 10.2 Foraminifera drawn by nineteenth-century naturalist Ernst Haeckel. Foraminifera are single-celled marine creatures that leave behind ornate shells. They can be used to track the history of nutrient pollution in estuaries. Public domain.

remains drop to the bottom, where microbes use up the available oxygen in breaking down their remains. If enough dead phytoplankton drop to the bottom layer, it will become hypoxic. When dissolved oxygen concentration drops below 2 mg per liter, every living thing that can swim away—fish, crabs, shrimp—will flee in search of oxygenated waters.

"What really gets hit," says Robert Diaz, a professor emeritus at the Virginia Institute of Marine Sciences, "are the benthic invertebrates." These creatures live burrowed into or physically attached to the bottom, and when the oxygen is gone they are doomed.

Kersey Sturdivant, a former student of Diaz who's now at Duke University, developed a camera system that showed impacts of hypoxia on Chesapeake Bay benthos in real time. As oxygen levels decreased, the chemical state of the bottom sediments changed, turning the mud from red to gray and black. Clams closed their shells, sealing themselves off from their surroundings. Worms that live burrowed into the sediment emerged and waved in the water, searching for oxygen. When oxygen plummets to zero, these creatures die. Anaerobic bacteria take over the bottom, metabolizing manganese sulfate from the sediment into poisonous hydrogen sulfide, H_2S, which tends to kill off the few benthic creatures that can survive without oxygen.

Sturdivant's studies of the benthos in Virginia's Rappahanock River, a tributary of the Chesapeake that goes hypoxic every summer, found that biological productivity drops by as much as 85 percent at low-oxygen sites.[31] Benthic creatures form a major part of the food chain for fish and crabs in the bay, and losing large areas of habitat and potential prey every summer likely impacts striped bass, blue crabs, and other creatures valued by Chesapeake watermen.

It can be difficult, however, to sort out the effects of hypoxia from those of over-harvest. In the 1980s, Chesapeake Bay's striped bass fishery collapsed. After a fishing moratorium was imposed, the striped bass population recovered during 1985–1995, years of intense summertime hypoxia in the mainstem bay.[32] A study of the distributions of bass and small prey fish during Chesapeake summers suggests that striped bass thrive when smaller fish are cornered in small pockets of oxygenated water, making them easy prey. But the bass population may now have reached numbers that can't be sustained by the available prey fish, which would account for a decline in striped bass numbers since 1995.

A study of long-term hypoxia in the Chesapeake found that below a threshold of 4.5 mg dissolved oxygen per liter, benthic diversity and biomass fall. Clams and oxygen-loving marine worms disappear, leaving one or two hardy kinds of polychaete worms that can survive on low oxygen. This means there is poor forage available for popular game fish like Atlantic croaker.[33] The loss of productive habitat likely depresses fish numbers, despite their ability to swim out of the dead zone.

Fish exposed to hypoxia can survive by swimming away—but they can end up impaired in serious but subtle ways. In the northern Gulf of Mexico, Atlantic croaker living in oxygen-deprived waters show impaired sexual development and produce greatly reduced numbers of sperm and eggs.[34] In nearby waters where oxygen is abundant, the sex ratio among croakers is about 50:50, but the majority of croaker in the dead zone are male. Gender is flexible among fishes, and environmental triggers can determine whether an individual becomes male or female. Hypoxia acts on genes that regulate sex determination, leading to generations of fish dominated by males. Over several generations, this pattern may threaten the survival of a population.[35]

If the pattern of nutrient overload, algal blooms, and oxygen depletion in bottom waters is the same worldwide, the life-and-death details can play out differently in each ecosystem. Which marine creatures die off has a great effect on how much a human-induced dead zone will capture human attention.

In the Kattegat, a shallow sea off the western coasts of Denmark and Sweden, a summertime dead zone built up in the 1970s. In 1975, fishers going after profitable Norway lobster had a boom year. The catch per unit effort more than doubled—but it happened because the lobsters were suffocating. Norway lobsters live in burrows in the bottom sediments. When oxygen concentrations plummet, they emerge from their burrows, making them easy targets for trawling fishermen.[36]

In 1986, more than half the lobsters caught in fisher's trawl nets were already dead or dying, victims of hypoxia. By 1988, the lobster population had collapsed. Underwater meadows of seagrass began to die off, deprived of sunlight by epiphytic algae that colonize blades of seagrass in nutrient-laden waters. Galvanized by the dramatic decline in the health of the Kattegat, the Danish Society for the Conservation of Nature lobbied the government to create policies that would curtail nutrient pollution. Upgrades in sewage treatment worked to reduce loads of nitrogen and phosphorus to some extent, but controlling nutrients in farm runoff was a tougher proposition.

The Danish government eventually put farmers on a nutrient diet, allowing them to apply only 90 percent of the fertilizer needed to produce maximum yields. Regulators tracked farmers' use of fertilizer, and penalized those who broke the rules by withholding subsidies. Denmark also initiated an ambitious plan to restore 16,000 hectares of wetlands to trap nitrogen-tainted farm runoff.[37] Less than twenty years after Denmark began to budget farm nutrients, the country reported a 50 percent drop in nitrogen and close to a 90 percent drop in phosphorus loads in its rivers and coastal seas.

Phytoplankton populations have dropped in parallel with the decline in nutrient loads. The aquatic plants of the seafloor have begun to recover, and seagrass meadows are recolonizing lost ground. But the bottom waters remain depleted of oxygen in summer, and water clarity has been slow to improve.

Scientific understanding of the damage done by eutrophication has grown tremendously in recent decades, but the process of ecosystem recovery remains little known. A eutrophic system can flip into an altered but stubbornly stable state. In the case of the Kattegat, climate change seems to have contributed.

Oxygen depletion happens in waters that are stratified. The tendency toward stratification has increased as ocean temperatures rise and winds that might mix the water column decrease. In the 1990s, the Kattegat seafloor held dense reefs of clams and mussels, filter feeders that scour phytoplankton and suspended particles out of the water. When the water column mixes freely, a healthy population of bivalves can filter the entire volume of an estuary several times a day. Once stratification sets in, filter feeders are cut off from a large chunk of the ecosystem and are also vulnerable to hypoxia.[38] Since 1990, even as Denmark worked to remedy its nutrient pollution problem, the population of filter feeders in the

Kattegat declined. Their loss will slow the ecosystem's healing. The complex community may never reassemble in its original state.

The most dramatic story of eutrophication and recovery comes from the northwestern Black Sea, where the River Danube carries the drainage from watersheds spread across eleven central European nations. The loads of nitrogen and phosphorus there doubled between 1960 and the 1980s, a time when the Soviet Union subsidized synthetic fertilizer use on row crops and the creation of giant livestock farms that released heavy loads of manure.

In the 1970s and 1980s, masses of dead fish and rotting crabs, clams, and mussels began to wash up on Black Sea beaches in Romania and Ukraine.[39] At its peak in 1990, the Black Sea's dead zone covered 40,000 square kilometers. The seafloor in this vast patch, once dense with filter-feeding bivalves and benthic algae, became barren. Long-time fisheries collapsed.[40] The ecosystem's recovery began in 1989 with the demise of the former Soviet Union. Government subsidies for agriculture disappeared, and the farming economy in the region fell apart. During the 1990s, nutrient loads diminished and HABs faded. Native zooplankton and fish species returned.[41] Conservation groups in the watershed worked to limit nutrient pollution as the agricultural economy recovered.

For a short time, the Black Sea presented a hopeful example of an ecological disaster repaired. Today, however, the northwestern Black Sea once again suffers from serious pollution, and marine life is dying. War in the watershed makes conservation work difficult or impossible.[42] The fate of coastal waters is inevitably tied to human politics onshore.

Perhaps the ultimate test of humanity's ability to clean up its nutrient-polluting act lies in the Mississippi River Basin. Forty-five percent of the land in the continental US drains into the Mississippi, and much of it is intensely farmed. The earth is veined with tile drains and saturated with nitrogen from decades of growing row crops dependent on synthetic fertilizer. The nutrient overload the Mississippi delivers to the northern Gulf of Mexico has created the largest coastal dead zone in the Western Hemisphere.[43]

Worldwide, the use of synthetic nitrogen fertilizers has increased more than eight-fold since 1960.[44] Manufactured nitrogen exceeds all other sources of bioavailable nitrogen combined. One-third of humanity's protein supply now relies on synthetic fertilizers.

The Midwest has come to specialize in growing corn and soybeans, nutrient-hungry crops that leak prodigious amounts of nitrogen and phosphorus to surrounding streams. There's a direct link between the percentage of a watershed area planted in these row crops and the concentration of nitrogen flowing out of that watershed.[45]

"You can predict stream nitrate by how much row crop acreage is in a watershed," says Keith Schilling, a research scientist at the University of Iowa. "Where you have high row crop intensity, you're likely to have more livestock, greater tile drainage, greater sediment and phosphorus loss."

Federal and state subsidies encourage both the production of synthetic nitrogen fertilizers and the farming of corn. The Energy Independence and Security Act of 2007, which mandated mass production of corn ethanol to encourage energy independence in the US, led to a dramatic rise in the price of corn. The natural result was that farmers took marginal lands that had been set aside as prairie or wetland under the US Department of Agriculture (USDA)'s Conservation Reserve Program and planted them in corn, accelerating nitrogen pollution.

Eileen McLellan, a senior scientist with the Environmental Defense Fund, works to find nitrogen hotspots in the Mississippi Basin, places where a strategically placed wetland or a stretch of floodplain restored in a tributary stream can capture significant amounts of pollution.

"Any strategy to deal with eutrophication that depends upon almost universal adoption of practices is doomed to failure," explains McLellan. She cites the example of cover crops, which can be grown when a field would otherwise lie fallow and dramatically reduce the release of nitrogen. Currently, about 3 percent of Midwestern farmers use cover crops—so the practice isn't making a dent in the overall problem. Instead of hoping for widespread changes that may never come, she's working toward small projects that can have a major impact.

Using a model of nitrogen pollution in the Mississippi watershed created by the US Geological Survey, McLellan pinpoints areas that release heavy loads of nitrogen. Then she and her colleagues meet with farmers in those regions and present a number of possible options that could curb pollution. These include creation of wetlands as well as the restoration of natural floodplains. Most Midwestern streams have been channelized. Flattening out the banks of stream so that they flood during times of high flow creates habitat where nitrogen is retained and processed, much as it is in wetlands.

"We'll use planning tools to identify maybe 100 different opportunities across a small watershed, where floodplain restoration or a restored wetland could be created. We lay all that out on a map, and hope to find twenty farmers willing to do a project on their land," explains McLellan.

During a recent study funded by the USDA, McLellan spoke with farmers in three different Mississippi Valley communities. "Under the Clean Water Act, there's no legal requirement for farmers to address nonpoint pollution," she notes. "So I always imagined that the first comment we'd hear would be on the lines of 'forget it, we don't have to do anything.' But that never came up."

The farmers McLellan worked with recognize that sooner or later, they'll be held accountable for the nutrient pollution that comes with the current cropping system. "They're open to looking at all the possible tools that will help manage pollution without putting them out of business," she says. "That's even more true since Des Moines Waterworks filed its lawsuit."

The city of Des Moines, Iowa, draws its drinking water from the Raccoon and Des Moines rivers. The city's waterworks, responsible for providing drinking water to 500,000 people, lies downstream of one of the most intense nitrogen pollution hotspots in the world. The Raccoon River drains 2.3 million acres of row-cropped Iowa farmland. The watershed is also home to a number of Confined Animal

Feeding Operations (CAFOs), which raise millions of hogs. Hogs far outnumber humans in Iowa, and waste from hog CAFOs is released to the state's streams, adding to the nutrient load. Iowa occupies less than 5 percent of the Mississippi Basin but is the source of 25 percent of the nitrogen that flows downriver to the Gulf of Mexico. Most of the nitrogen flows downstream in the form of nitrate, NO_3. Nitrate can be toxic, especially to infants, so the EPA has established a limit of 10 mg nitrate per liter of drinking water. Concentrations of nitrate in Iowa streams and rivers often far exceed that limit.

In 1993, Des Moines Waterworks finished building the world's largest nitrate removal facility, a plant that pulls nitrate out of the water in an expensive, labor- and energy-intensive process. The nitrate removal plant was set up to run on an as-needed basis; it costs about $7,000 per day to run, so it was fired up only when nitrate levels in the incoming river water approached the limit of 10 mg per liter. But over time, the nitrate loads in the rivers rose and the plant had to operate more often. The summer of 2013 saw peak nitrogen loads in the Raccoon River; that year Des Moines Waterworks spent $900,000 on nitrate removal, necessitating a rate hike for its customers. When peak nitrogen loads appeared again in the summers of 2014 and 2015, managers at Des Moines Waterworks realized that the nitrate removal plant would soon need to be rebuilt and considerably enlarged, at a cost in the neighborhood of $100 million. Twenty years after it was built, the world's largest nitrate removal plant had been made obsolete by the prodigious pollution from Iowa agriculture.[46]

In 2015, Des Moines Waterworks filed a federal lawsuit against three upstream counties, demanding that they control the nutrient pollution from their drainage districts. The lawsuit, the brainchild of Waterworks CEO Bill Stowe, asks for radical changes in the way the CWA has traditionally been applied to agricultural pollution. The suit argues that the outflow from drainage districts should be regulated as point sources of pollution, just as an industrial discharge or a city sewage plant would. Focused on local problems, the suit raises serious questions of national policy.

The legal move by the Waterworks is controversial. Stowe has been criticized by both Republicans and Democrats who feel he should have allowed time for the state's voluntary plan to address agricultural pollution, the Iowa Nutrient Reduction Strategy, introduced in 2013, to work. The Iowa Partnership for Clean Water, a bipartisan group backed by the Iowa Farm Bureau, has run ads accusing Stowe of wasting hundreds of thousands of dollars on an "outrageous" lawsuit targeting farmers.[47]

Iowa's Nutrient Reduction Strategy acknowledges that currently unregulated sources create 92 percent of the state's nitrate pollution. The remedies suggested in the Strategy are all voluntary. The state offers farmers financial incentives for conservation actions that will reduce nitrate leakage, but the plan includes no timeframes or numerical criteria for pollution reduction. In Stowe's eyes, this approach is far too passive to address a widespread environmental crisis.

"The question," he says, "is how are we ever going to get industrial agriculture to rein in their practices so there is less public health risk to those of us downstream. That's what our lawsuit is about. Our state gives economic incentives to manufacturers of synthetic nitrogen fertilizer, and to animal feeding operations that produce more nitrogen pollution than all of Iowa's row crops. The agricultural model is broken."

Industrial corn and soybean farming feeds a market for low-cost food, but as it's currently practiced it precludes a real improvement in water quality. "It's your classic wicked problem," says Schilling. "The nutrient reduction strategies just work around the edges a bit. If absolutely everybody reduces their fertilizer use to the minimum, that gets you about a 10 percent reduction in nitrogen pollution." But 10 percent is not nearly enough to make Raccoon River water safe for drinking, or to begin to heal the dead zone in the Gulf.

The Midwestern focus on corn is traditional, and it's difficult to get farmers to change. Government subsidies encourage the existing system of nutrient-leaky row crops and intense hog-feeding operations. There are alternatives: Corn could be rotated with other grains, or with alfalfa. In terms of biofuel, switchgrass, a native perennial prairie grass, makes an excellent source, using a fraction of the fertilizer and requiring none of the tile drainage involved in growing corn. But a crop of switchgrass can take three years to mature, and the government isn't subsidizing the process. Most existing biofuel factories are set up to handle corn, not switchgrass. As things now stand, even the most conservation-minded farmer can't turn to switchgrass as a cash crop.

Despite its high levels of nitrate pollution, Iowa is a national leader in creating wetlands to capture nutrients in farm runoff. The state has been designing wetlands placed to capture runoff from small sub-watersheds since 2001, based on years of research by wetlands ecologist William Crumpton of Iowa State University. Shawn Richmond, now the Environmental Technology Director for the Agribusiness Association of Iowa, studied under Crumpton and has helped design and build dozens of wetlands.

"We target the heavily tile-drained parts of the state," explains Richmond. Using computerized mapping and analysis of a watershed, he searches for places in the landscape where an effective wetland can fit. An ideal location is at the base of a watershed, where natural topography creates a dip that can be transformed into a marsh by adding an earthen berm. Once researchers have found a good place for a wetland, they get on the phone to the person who owns the land. "It's cold-call conservation," says Richmond. "About one in three landowners will show interest when we call out of the blue."

The average wetland project takes up about forty acres, ten acres in wetland surrounded by thirty acres of native prairie plants. A strategically placed wetland can capture up to 70 percent of the nitrate flowing out of a sub-watershed, using only a small fraction of the total cropland. On the other hand, a farmer who gives up forty acres is making a real sacrifice.

"That's better than $250,000 worth of land taken out of production," notes Richmond. State agencies pay the cost of creating the wetland, and offer soil rental rates and a lump-sum payment to farmers who commit to keeping the wetland in perpetuity.

Iowa has about eighty of these nutrient-capture wetlands built. They're more effective than the few that have been created in Ohio, where the location of wetlands is based on the inclination of landowners, not on optimal positioning in the landscape. State biologists in Ohio often have to run the tile drains of neighboring farmers under or around a constructed wetland, minimizing its ability to capture and filter polluted runoff. While Iowa's progress on this front is impressive, there's a long way left to go.

"In Iowa, we're going to need 4,000–7,000 constructed wetlands, in addition to other conservation practices," says Richmond. Another important strategy is the construction of bioreactors, small trenches dug across the outflow of a field tile and filled with wood chips. The wood chips act as a source of carbon, essential to the process of denitrification, carried out by soil microbes. Richmond estimates the state will need about 120,000 bioreactors to bring its nitrogen pollution problem under control.

The bottleneck in Iowa's progress is funding. The state legislature appropriates about $1 million per year for wetland restoration on farmland, enough to pay for three or four projects a year.

"The science is pretty far along in terms of identifying the best locations for potential wetlands," says Schilling. "It's a question of having the money, the resources, and the will to do it."

If you do the math, and bear in mind that Iowa is in some ways more progressive than other Corn Belt states whose fields drain to the Mississippi, the future looks grim. In terms of changing the way people manage the watershed, progress will be slow, and the incentives in current federal policies are backwards.

Nancy Rabalais, a marine ecologist at the Louisiana Universities Marine Consortium, has monitored the dead zone in the Gulf of Mexico since 1985. Remains of decades-old foraminifera show that hypoxia built in the northern Gulf beginning in the 1950s, accelerated for decades, and has reached a plateau in recent years. "Healing the dead zone is going to take a societal shift in the way we farm," she says. "There are sustainable, ecologically sound ways to farm that cost less than what's done now. It's just a mindset of how agriculture is done."

The landscape itself is also likely to resist an easy shift toward lower nutrient levels. Climate models predict more frequent and intense rainstorms in the Midwest, which will trigger more nutrient-laden runoff. The vast Mississippi watershed has already built up a legacy of excess nutrients that may last for decades after ongoing pollution is controlled.

"Nitrogen fertilizer that was applied decades ago has made its way down into the groundwater," explains McLellan. "That groundwater will eventually make its way to the streams." This is why an ability to process nitrogen and pump it back into the air via denitrification must be restored to creeks and rivers, through restoration of floodplains, backwaters, and oxbows long-ago lost to our passion for levees and straight rivers.

The modern problem of nutrient pollution is the result of intense human tinkering with the flow of water and the natural cycling of nitrogen. The settlement of the US by a people obsessed with draining and destroying wetlands came first; the twentieth-century discovery of a way to fix nitrogen on an industrial scale sealed the fate of our coastal waters. Yet sometimes, researchers studying the deterioration of aquatic life stumble on a revival that seems to have come out of nowhere. Off the nutrient-enriched coasts of the US and Europe, the return of top predators may help rescue smothered ecosystems.

NOTES

1. Fitzsimmons, E. (August 3, 2014). "Tap Water Ban for Toledo Residents." *New York Times* (http://www.nytimes.com/2014/08/04/us/toledo-faces-second-day-of-water-ban.html?_r=0); Henry, T. (August 3, 2014). "Water crisis grips hundreds of thousands in Toledo area, state of emergency declared." *Toledo Blade* (http://www.toledoblade.com/local/2014/08/03/Water-crisis-grips-area.html).
2. Harke, M., Morgan Steffen, Christopher Gobler, Timothy Otten, Steven Wilhelm, Susanna Wood, Hans Paerl (2016). "A review of the global ecology, genomics and biogeography of the toxic cyanobacterium, Microcystis spp." *Harmful Algae* **54**: 4–20.
3. Watson, S., Carol Miller, George Arhonditsis, et al. (2016). "The re-eutrophication of Lake Erie: Harmful algal blooms and hypoxia." *Harmful Algae* **56**: 44–66.
4. Michalak, A., Eric Anderson, Dmitry Beletsky, et al. (2013). "Record-setting algal bloom in Lake Erie caused by agricultural and meteorological trends consistent with expected future conditions." *PNAS* **110**(16): 6448–6452.
5. Harke, M., Morgan Steffen, Christopher Gobler, Timothy Otten, Steven Wilhelm, Susanna Wood, Hans Paerl (2016). "A review of the global ecology, genomics and biogeography of the toxic cyanobacterium, Microcystis spp." *Harmful Algae* **54**: 4–20.
6. Kaatz, M.R. (1955). "The Black Swamp: a study in historical geography." *Annals of the Association of American Geographers* **XLV**(1): 1–36.
7. Scavia, D., J. david Allan, Kristin K. Arend, et al. (2014). "Assessing and addressing the re-eutrophication of Lake Erie: Central basin hypoxia." *Journal of Great Lakes Research* **40**: 226–246.
8. Mitsch, W. (2015). "Restoring the Great Black Swamp to save Lake Erie." https://www.linkedin.com/pulse/restoring-black-swamp-save-lake-erie-bill-mitsch.
9. Mitsch, William, James Gosselink (2015). "Wetlands." Hoboken, New Jersey: Wiley.
10. Odum, H. T., K.C. Ewel, W.J. Mitsch, J.W. Ordway (1977). "Recycling treated sewage through cypress wetlands in Florida." In "Wastewater renovation and reuse," Frank M. D'Itri editor, *Marcel Dekker, Inc.*
11. Baker, D. (2010). "Trends in bioavailable phosphorus loading to Lake Erie." *Lake Erie Protection Fund Grant 315-07.*
12. Ohio Environmental Protection Agency (2010). "Ohio Lake Erie Phosphorus Task Force Final Report: Executive summary."
13. Watson, S., Carol Miller, George Arhonditsis, et al. (2016). "The re-eutrophication of Lake Erie: Harmful algal blooms and hypoxia." *Harmful Algae* **56**: 44–66.
14. Michalak, A., Eric Anderson, Dmitry Beletsky, et al. (2013). "Record-setting algal bloom in Lake Erie caused by agricultural and meteorological trends consistent with expected future conditions." *PNAS* **110**(16): 6448–6452.

15. Lenhart, C., Peter Lenhart (2014). "Restoration of wetland and prairie on farmland in the former Great Black Swamp of Ohio, USA." *Restoration Practice* **32**(4): 441–449.
16. King, K., Mark Williams, Merrin Macrae, et al. (2015). "Phosphorus transport in agricultural subsurface drainage: a review." *Journal of Environmental Quality* **44**: 467–485.
17. Scavia, D., Margaret Kalcic, Rebecca Logsdon Muenich, et al. (2016). "Informing Lake Erie Agriculture Nutrient Management via Scenario Evaluation." *University of Michigan Water Center* (http://graham.umich.edu/water).
18. NOAA (2017). "Gulf of Mexico dead zone is the largest ever measured." http://www.noaa.gov/media-release/gulf-of-mexico-dead-zone-is-largest-ever-measured.
19. Sklar, F., Michael Chimney, Susan Newman, et al. (2005). "The ecological-societal underpinnings of Everglades restoration." *Frontiers in Ecology and Environment* **3**(3): 161–169.
20. Rizzardi, K.W. (2001). "Alligators and litigators: a recent history of Everglades regulation and litigation." *Florida Bar Journal* **75**(3): 18–27.
21. Kadlec, R.H. (2006). "Free surface wetlands for phosphorus removal: the position of the Everglades Nutrient Removal Project." *Ecological Engineering* **27**: 361–379.
22. Levy, S. (August 2015). "The ecology of artificial wetlands." *BioScience* 65(4): 346-352): 1–7.
23. Kadlec, R.H., D.L. Hey (1994). "Constructed wetlands for river water quality improvement." *Water Science and Technology* **29**(4): 159–168; Kadlec, R.H. (2009). "The Houghton Lake wetland treatment project." *Ecological Engineering* **35**: 1285–1286.
24. Henry, T. (July 24, 2016). "Lake Erie, South Florida algae crises share common toxins and causes." *Toledo Blade*.
25. Hansson, A., Eja Pedersen, Stefan Weisner (2012). "Landowners' incentives for constructing wetlands in an agricultural area in south Sweden." *Journal of Environmental Management* **113**: 271–278.
26. Diaz, R., Rutger Rosenberg (2008). "Spreading dead zones and consequences for marine ecosystems." *Science* **321**(5891): 926–929.
27. Galloway, J., Alan Townsend, Jan Willem Erisman, et al. (2008) "Transformation of the Nitrogen cycle: recent trends, questions and potential solutions." *Science* **320**: 889–892.
28. Mee, L. (November 2006). "Reviving dead zones." *Scientific American*, pp. 79–85.
29. Rabalais, N., R. Eugene Turner, William J. Wiseman, Jr. (2002). "Gulf of Mexico hypoxia, a.k.a. the dead zone." *Annual Review of Ecology and Systematics* **33**: 235–263.
30. Rabalais, N., R. Eugene Turner, Barun K. Sen Gupta, et al. (2007). "Sediments tell the history of eutrophication and hypoxia in the northern Gulf of Mexico." *Ecological Applications* **17**(5 Supplement): S129–S143.
31. Sturdivant, S.K., Rochelle Seitz, Robert Diaz (2013). "Effects of seasonal hypoxia on macrobenthic production and function in the Rappahannock River, Virginia, USA." *Marine Ecology Progress Series* **490**: 53–68.
32. Constantini, M., Stuart Ludsin, Doran Mason, et al. (2008). "Effect of hypoxia on habitat quality of striped bass (*Morone saxatilis*) in Chesapeake Bay." *Canadian Journal of Fisheries and Aquatic Science* **65**: 989–1002.
33. Seitz, R., Daniel Dauer, Roberto Llanso, W. Christopher Long (2009). "Broad-scale effects of hypoxia on benthic community structure in Chesapeake Bay, USA." *Journal of Experimental Marine Biology and Ecology* **381**: S4–S12.

34. Thomas, P., M. Saydur Rahman (2010). "Region-wide impairment of Atlantic Croaker testicular development and sperm production in the northern Gulf of Mexico hypoxic dead zone." *Marine Environmental Research* **69**: 559–562.
35. Cheung, C.H.Y., Jill Man Ying Chiu, Rudolf Shiu Sun Wu (2014). "Hypoxia turns genotypic female medaka fish into phenotypic males." *Ecotoxicology* **23**: 1260–1269.
36. Baden, S., Lars-Ove Loo, Leif Pihl, Rutger Rosenberg (1990). "Effects of eutrophication on benthic communities including fish: Swedish west coast." *Ambio* **19**(3): 113–122.
37. Kronvang, B., Hans Andersen, Christen Borgesen, et al. (2008). "Effects of policy measures implemented in Denmark on nitrogen pollution of the aquatic environment." *Environmental Science and Policy* **11**: 144–152.
38. Riemann, B., Jacob Carstensen, Karsten Dahl, et al. (2016). "Recovery of Danish coastal ecosystems after reductions in nutrient loading: a holistic ecosystem approach." *Estuaries and Coasts* **39**: 82–97.
39. Mee, L. (November 2006). "Reviving dead zones." *Scientific American*, pp. 79–85.
40. Schwartzstein, P. (May 11, 2016). "The Black Sea is dying, and war might push it over the edge." *Smithsonian* (http://www.smithsonianmag.com/science-nature/black-sea-dying-and-war-might-push-it-over-edge-180959053/).
41. Kideys, A.E. (2002). "Fall and rise of the Black Sea ecosystem." *Science* **297**(5586): 1482–1484.
42. Schwartzstein, P. (May 11, 2016). "The Black Sea is dying, and war might push it over the edge." *Smithsonian* (http://www.smithsonianmag.com/science-nature/black-sea-dying-and-war-might-push-it-over-edge-180959053/).
43. Rabalais, N., R. Eugene Turner, William J. Wiseman, Jr. (2002). "Gulf of Mexico hypoxia, a.k.a. the dead zone." *Annual Review of Ecology and Systematics* **33**: 235–263.
44. Broussard, W., R. Eugene Turner (2009). "A century of land use and water quality relationships in the continental US." *Frontiers in Ecology and Environment* 7(6): 302–307.
45. Ibid.
46. *Des Moines Water Works vs. Sac County, Buena Vista County, Calhoun County.*" Case 5:15-cv-04020-DEO, US District Court for the Northern District of Iowa, Western Division, 2015.
47. Masters, C. (January 21, 2016). "Iowa's nasty water war." *Politico*.

11

Wild Things

A group of sea otters laze at the edge of Elkhorn Slough. They float on their backs in the steel-gray water, paws folded against their chests, gazing at the small boat steered by ecologist Brent Hughes of the University of California–Santa Cruz. Hughes has documented a profound shift in the slough's ecology, triggered by the otters. Sea otters were nearly driven to extinction by fur hunters in the 1800s, and were gone from Elkhorn Slough for a century. In 1984, when the first sea otters recolonized, Elkhorn Slough's once bountiful eelgrass beds had dwindled to a few small, scattered patches. Now, more than thirty years after the sea otters' return, expanding eelgrass beds grow lush beneath the water's surface, the dense leaves sheltering juvenile fish and feeding an array of invertebrate grazers.

The slough, on the central California coast, is one of the most severely polluted estuaries on the planet. Artificial fertilizer applied to 2.69 million acres of farmland in the neighboring Salinas Valley runs into its waters. The excess nutrient load causes eutrophication. It also fuels the growth of epiphytic algae that thrive on the surface of eelgrass leaves, blocking the sunlight the grass needs and smothering whole beds.

The problem is common in estuaries around the globe, which receive heavy loads of nutrients from rivers draining polluted watersheds. Seagrass meadows filter contaminants from water and prevent coastal erosion in addition to acting as nurseries for fish and invertebrates. These crucial habitats are disappearing. The global distribution of seagrasses has decreased by 29 percent over the last 140 years, and 58 percent of the surviving seagrass meadows are in decline.[1] Nutrient pollution of coastal waters had long been thought to be the main driver of this trend. But in Elkhorn Slough, the eelgrass has made a remarkable comeback even as pollution loads continued to climb. The mechanism of this welcome ecological shift was unknown until Hughes demonstrated that sea otters are the key (Fig. 11.1).

He began to put the pieces of the puzzle together when he went diving in Tomales Bay, an unpolluted estuary to the north. The eelgrass in Elkhorn Slough was lush and green despite intense pollution; in Tomales Bay, where there are no sea otters, the eelgrass was a dull brown, smothering under epiphytic algae. Hughes noticed another difference: The sea slugs, invertebrate grazers that feed

Figure 11.1 Sea otter in Morro Bay, California. Sea otters have been found to have a powerful positive impact in the highly polluted ecosystem of Elkhorn Slough. Photo by Mike Baird.

on algae, were much bigger and more abundant in Elkhorn Slough. "Here they're like mutant monsters," he says. He began to wonder if he was seeing the result of a trophic cascade—a chain reaction in which predatory sea otters benefited seagrasses.

Sea otters were thought to be gone from California until a small colony of survivors was discovered on the Big Sur coast in 1938. Under strict protection, the otter population has gradually expanded. When otters entered Elkhorn Slough in 1984, they found a bonanza of easy prey: abundant crabs. The crabs feed on invertebrate grazers like sea slugs and amphipods (shrimp-like crustaceans), which devour the epiphytic algae that can otherwise smother eelgrass. During the otters' long absence, crabs had become dominant and grazers declined, allowing epiphytic algae to thrive in the polluted water.

"The sea otter is a model top predator," notes Hughes. "They eat 25 percent of their body weight daily." Otters need the calories because they lack the blubber that insulates other marine mammals—and due to their voracious appetites, otters can have a major impact on the numbers of their prey. They make very convenient study subjects, because they bring their food to the surface where it can easily be observed. An army of citizen volunteers has been tracking the eating habits of the otters since they returned to Elkhorn Slough; the data show they feed heavily on crabs. When Hughes looked up historical data on otter populations and the expansion of eelgrass beds in the slough, "the pattern matched like a hand in a glove," he says.[2] He and his colleagues confirmed the trophic cascade hypothesis using laboratory experiments and predator-exclusion cages in parts of the slough.

Hughes was surprised by his discovery. He'd spent much of his graduate education studying algae and had been mentored by scientists who believed that bottom-up forces, like changes in temperature and nutrient concentrations, were the principal explanation for the global decline of seagrasses. Because it wasn't what he was primed to see, "it took a lot to recognize the impact of otters," he says. The trophic cascade in Elkhorn Slough is a classic case of top-down regulation of an ecosystem—where the balance is shifted by interactions among predators, grazers, and plants.

The otter population in the slough continues to climb. "We keep thinking the slough is at carrying capacity," says Hughes, "but the otters keep surprising us." Every eelgrass bed in the slough continues to expand as otter numbers grow. When Hughes and his colleagues use cages to exclude otters from an area, the eelgrass inside the cage gathers a load of epiphytic algae and declines.

During the past five years, otters began to haul out in salt marshes at the slough's edge. Mother otters find the marshes particularly useful as a safe place to leave their pups while they hunt. Few predators capable of taking a pup lurk in the shallow waters of the marsh. There are now generations of otters in the slough who've never swum in the open ocean, once thought to be the species' only natural habitat. Though its waters are murky with human-generated pollutants, Elkhorn Slough now offers a window into the ways of sea otters before European settlement.

A parallel case, illustrating the role of top predators in protecting seagrass beds, has been documented at the edge of the North Sea, on the west coast of Sweden. There, nutrient pollution has steadily increased at the same time that overfishing has led to a steep decline in the population of adult cod, once the major marine predator in the region.[3] More than 58 percent of local seagrass beds (*Zostera marina*, the same species that grows in Elkhorn Slough) have been lost since the 1980s.[4]

On Sweden's Baltic coast, the water is also polluted and eutrophic, and cod stocks have been depleted over the past twenty-five years. Yet there has been no significant decline in seagrass beds. The critical difference lies in the food web of each region. Sweden's North Sea coast hosts an abundant array of small predators, including grass shrimp, gobid fish, and shore crabs, which feed on grazers, releasing epiphytic algae to thrive and smother the eelgrass. In the Baltic, small predators are scarce, grazers are abundant, and the eelgrass survives.[5] Experiments with fertilization of eelgrass beds show that nutrient loading does not cause a bloom of epiphytic algae where grazers are present. Instead, grazers become more abundant, devouring the algae as fast as it can grow.

The results of studies in eelgrass beds in Chesapeake Bay, led by Emmett Duffy of the Virginia Institute of Marine Sciences, echo the Swedish findings. In some experimental plots the number of amphipods, small crustaceans that are dominant grazers on epiphytic algae, was lowered by using carbaryl insecticide. In others, artificial fertilizer was applied. Added nutrients caused a six-fold increase in growth of epiphytic algae, reducing eelgrass growth by 65 percent.[6] Where natural grazer populations were present, they controlled algal growth, and fertilization actually increased eelgrass biomass. The presence of small predators, including fish and crabs, lowered populations of grazers, increased biomass of algae, and inhibited growth of eelgrass.

"What we found is similar to what's been shown in Elkhorn Slough," says Duffy. "The system we looked at had three levels: small predators, grazers, and algae. Hughes went up one more level, with the addition of otters as an apex predator. Depending on the structure of the food chain, the presence or absence of predators may have different impacts on the plants."

Elkhorn Slough presented a rare natural experiment in the return of a lost top predator; the situation in the North Sea, with the decline of cod at the same time that nutrient pollution increased, is a clear parallel. In many other ecosystems, it's more difficult to nail down the shifting balance among predators, prey, and the plants that form the foundation of ecosystems. The Chesapeake, for instance, hosts multiple species of predatory fish, many affected by overfishing, but detailed data on their role in the eelgrass ecosystem don't exist.

Duffy is lead researcher on an international project studying bottom-up and top-down forces affecting eelgrass throughout its range. The Zostera Experimental Network (ZEN) is conducting standardized experiments in eelgrass beds at fifteen sites scattered across the Northern Hemisphere. Researchers tested the effects of nutrient fertilization and of reducing the population of crustacean grazers. The results surprised Duffy.

"We found no effect of nutrient loading alone on the amount of algae," he says. "We did see bottom-up control of amphipods, the small animals that eat the algae. They're more abundant in areas with higher nitrogen loads. That means you do have bottom-up control, but the increase in biomass passes through the algae, and is realized at the level of the animals feeding on it."

The ZEN study also examined genetic diversity within eelgrass beds and species richness of grazers. More genetically diverse eelgrass hosted more small grazers, and more diverse grazer communities seem better able to control algal growth. The findings underscore the need to protect biodiversity in a human-dominated world.

"We've learned over the last two decades that changes in top predator populations, on land and sea, can have a pervasive influence on the rest of the ecosystem," notes Duffy. "We want to have clear water and healthy seagrass beds and marshes. Large predators are very vulnerable to human impacts, and losing them can trigger the loss of these habitats."

The long-held assumption that coastal habitats are shaped only by bottom-up influences arose from classic studies of salt marshes on Sapelo Island, Georgia, done by Eugene Odum and his colleagues in the 1950s and 1960s. Odum focused on the role of physical factors, including temperature, salinity, and the availability of nutrients, in shaping marsh productivity, and saw grazers as having little effect. The idea that only bottom-up forces affect salt marsh ecosystems became a scientific dogma, and was applied to seagrass meadows and mangrove forests as well as marshes.[7]

The first evidence countering the bottom-up model came in the late 1980s, when salt marsh began to disappear from the shore of La Perouse Bay in Manitoba. Robert Jefferies of the University of Toronto documented the process, caused by grazing of large flocks of lesser snow geese on their breeding grounds. The geese grazed salt marsh plants to the ground, then grubbed into the soil to devour the roots, transforming the landscape to barren mud.[8] The habitat change has been longstanding, because evaporation increases in the open mudflat, making the environment too saline for marsh plants to recolonize. This marsh loss in subarctic Canada was driven by human-caused habitat change far to the south. While many of the temperate zone salt marshes historically used by migrating geese have disappeared as a result of coastal development, the lesser snow goose has flourished during its winter migration by feeding in agricultural fields and on golf courses, which are heavily fertilized.[9] Brian Silliman, now at Duke University, and his one-time advisor Mark Bertness, of Brown University, have pioneered studies that show the devastating impact of smaller, less obvious grazers on salt marshes—an effect that can be intensified by nutrient pollution. In one early experiment, Bertness and Silliman fertilized plots of salt marsh on Sapelo Island and compared the growth of areas with and without herbivorous snails. They found that fertilized marsh plants also attracted hungry insects. "About two years in, we went out to the marsh and saw that our experiment was over," remembers Bertness. "Every single one of the fertilized plots had been completely devoured by grasshoppers."

Over the last fifteen years, both Bertness and Silliman have documented a dramatic role of nutrient pollution in intensifying the impact of grazers on salt marsh. When nitrogen is abundant, marsh plants incorporate it into their leaves, producing soft, nutritious foliage that's irresistible to grazers. The researchers have witnessed an intense response from a variety of grazing animals, ranging from insects and snails to cattle in an ongoing study in Chile.

"Fertilize a spot on the marsh and the cattle are so intent on getting to that area that they'll knock cages over, stick their heads through barbed wire," says Bertness. "There's marsh grass all around, but they want the fertilized stuff."

The phenomenon appears to play an important role in the die-off of salt marshes, which has wiped out more than 250,000 hectares[10] of marsh habitat along the Atlantic and Gulf coasts of the US. "The heavy flow of nitrogen and phosphorus into these marshes from upstream cities and farms can trigger a chain reaction that leads to intense overgrazing by marsh herbivores," says Silliman. Native snails alone can devastate tracts of nutrient-enriched marsh plants if predators are not there to control their numbers.

Silliman first became interested in the role of marsh grazers when he noticed a native snail, the periwinkle, feeding on green stems of cordgrass in a Virginia salt marsh. Periwinkles feed mainly on standing dead cordgrass and detritus, and so had been thought to have little influence on cordgrass growth. Silliman discovered that by rasping away at live shoots of grass, the periwinkles created and maintained wounds that were invaded by fungi. In effect, the snails injure the cordgrass in order to farm these edible fungi. In 2002, Silliman published a study that tracked a wave of snails as it denuded a once-healthy tract of cordgrass in only eight months. The periwinkle's major predators—the blue crab and the diamondback terrapin—had been hunted into scarcity, allowing the snail population to expand. "The plants have never seen the kind of increases in nitrogen we're giving them now," says Silliman, whose work has focused on marshes in the southeastern US. Coastal woodlands once absorbed much of the excess nitrogen, but as development accelerates, the woods are cut down and nitrogen availability in marshes skyrockets, increasing by 200 to 300 percent.

Seagrass meadows and salt marshes are among the most effective natural filters for nutrient pollution. The thick growth of plants in these habitats slows the flow of water, causing organic material to settle to the bottom, where some of the carbon and nitrogen becomes locked in the sediment. They also have an unparalleled ability to remove nitrogen, the main nutrient of concern in marine waters, from the ecosystem. Denitrification, the bacterial process that transforms biologically available nitrate into N_2 gas, removes large amounts of nitrogen from coastal ecosystems. Work by Michael Piehler of the University of North Carolina has shown that more than 75 percent of the nitrogen load carried to Bogue Sound, North Carolina, by polluted streams is released to the atmosphere through denitrification. The process occurs at much higher rates among structured coastal habitats, like seagrass beds, salt marshes, and oyster reefs, than in barren mud.[11] Even small strips of remnant salt marsh can make a real improvement in denitrification rates.[12]

The runaway pollution caused by use of artificial nitrogen fertilizers threatens coastal ecosystems around the world. The array of life supported by dwindling marshes and seagrass beds is under serious threat. That reality is clear in Elkhorn Slough, where the return of lush eelgrass has failed to guard against other impacts of extreme nutrient pollution. In a recently published study, Hughes and his colleagues found that low levels of dissolved oxygen make the slough toxic to young fish, impairing its historical function as a nursery for marine life.[13] The result has been a significant decline in the population of commercially fished species, including English sole, in the open waters outside the slough.

"We need less nutrients and more intact top predator populations," says Hughes. "Given the intensity of human land use and population growth in many coastal watersheds, reducing nutrient pollution will be difficult. Restoring food webs may be the key to ecosystem survival."

At a lush south Florida marsh, an alligator lunges off the bank, triggering a mighty splash that sends a startled heron into flight. Mobs of white ibis and roseate spoonbill pluck their prey from the water or sift it from the mud. Brian Garrett, wildlife coordinator for the South Florida Water Management District (SFWMD), steps on the brakes as a handful of wood storks flush away from his pickup truck. The birds' lanky, bald-headed bodies transform to soaring grace as they circle above us.

This industrial habitat, created to filter pollutants out of the runoff from sugar and dairy farms upstream, is one of the Stormwater Treatment Areas (STAs) managed by the South Florida Water Management District (SFWMD). (See Chapter 10.) A recent study found that bird populations in the 57,000 acres of STAs are more abundant and diverse than those in found in adjacent natural marsh.[14]

The findings are part of a new wave of research on the wildlife ecology of constructed wetlands, critical habitats in a world where many natural wetlands have been destroyed. Thousands of manmade wetlands now treat sewage effluent and contaminated runoff from city streets and farm fields worldwide. These wetlands are most often designed to improve water quality, not to nurture wildlife. Yet creatures from coyotes to snails inevitably exploit the rich habitats they create. Depending on the circumstance and the species, constructed wetlands may represent a lifeline or an ecological trap—a place that lures adults to breed but is deadly to their young.

Tyler Beck, now a biologist with the Florida Wildlife Commission, studied bird populations in and around the STAs as a graduate student at Florida Atlantic University. He found that bird density in the artificial marshes was thirty-eight times greater than in nearby natural wetland, while species richness was four times higher. This result was no surprise to him after his first visits to the STAs. In addition to the relatively obvious birds, remote, motion-activated cameras have caught images of the endangered Florida panther and of abundant coyotes and feral hogs. Thousands of people come to the STAs each year to bird-watch, and hundreds more line up to hunt ducks and alligators. "Our study put down on paper what anybody can see with their own eyes out here," Beck says.

The STAs are part of a massive effort to restore the ecology of the Everglades, an expanse of sawgrass marsh, cypress swamp, and coastal mangrove forest that once encompassed 4.8 million acres in south Florida.[15] About 700,000 acres of the northern Everglades, at the edge of Lake Okeechobee, were drained to create farmland in a frenzy of canal-building that began in the early 1900s. Before this grand re-engineering of natural flows, a sheet of shallow, clear water ran south

Figure 11.2 American flamingoes at a Florida Stormwater Treatment Area. Photo courtesy South Florida Water Management District.

from Lake Okeechobee across the tip of the Florida peninsula, sustaining sawgrass marshes adapted to low-nutrient conditions—which the conservationist Marjory Stoneman Douglas famously named "the river of grass."

By the 1970s, Lake Okeechobee was choking on heavy loads of nutrients that ran off sugar fields and dairy farms. Blooms of algae fueled by high phosphorus levels have at times covered as much as 153 square miles—about twenty percent of the lake's surface. Such explosions of algal growth can pump toxins into the water, or drain it of oxygen, killing fish and aquatic invertebrates. The lake was suffering from a bad case of eutrophication. The polluted water flowed downstream into remaining native wetlands, creating an explosion of cattail, which thrives in nutrient-rich water, outcompeting sawgrass and reshaping the ancient Everglades ecosystem. The shift in vegetation affects everything from tiny invertebrates in the bottom muck to hawks cruising above the marshes, which have a hard time locating their prey amid the tall, dense stands of cattail.

In the late 1990s, as part of a settlement in a federal lawsuit over violation of water quality standards, SFWMD built six stormwater treatment areas at the interface between heavily fertilized agricultural land and protected native wetlands. Much of each STA consists of thick stands of cattail, which not only thrive in nutrient-laden water but excel at knocking down phosphorus levels. Cattail stems slow the flow of water, allowing phosphorus-rich sediment to settle to the bottom. The plants also take up nutrients to fuel their own growth. Farther downstream in each STA are areas kept clear of cattail so that submerged plants—like guppy grass (*Najas guadalupensis*) and muskgrass (*Chara* spp.) can flourish.[16] These underwater plants absorb more of the phosphorus load. The STA systems remove 60 to 80 percent of the total phosphorus, achieving a major improvement in water quality, though nutrient loads remain higher than those in the pristine, predevelopment Everglades.[17]

Bird populations are dense in the STAs because the nutrient-rich waters there fuel an intensely productive ecosystem, while the cattail stands filter out enough of the excess nutrients to prevent the worst symptoms of eutrophication (Fig. 11.2). American coots, small black-plumed, white-billed birds, form particularly abundant flocks. Beck calculated that 8 percent of the total American coot population is now wintering in the STAs—which implies that these artificial habitats may be altering the overall distribution of some North American water birds. The STAs have also become ground zero for the invasion of the purple swamphen, a showy species introduced from Asia, now flourishing alongside the native coots and gallinules.

The STAs support rare birds along with the common and the invasive. The snail kite, an endangered subspecies that lives only in Cuba and the Everglades region, has a sharply curved bill adapted to pry the flesh of native apple snails from their shells (Figs. 11.3 and 11.4). The kites prey almost exclusively on apple snails—including invasive species of apple snail introduced to Florida. They need relatively open wetlands, where they can see their prey as they cruise over the water. For this reason, stands of cattail make poor kite habitat.

Yet there's enough open water in the STAs that kites are a common sight, flying low with their heads tilted down as they search for prey. Beck, who is now the snail

Figure 11.3 A young snail kite. An endangered subspecies that lives only in Cuba and the Everglades region, the kite preys on apple snails. The birds breed successfully in the south Florida STAs. Photo courtesy South Florida Water Management District.

kite conservation coordinator for the Florida Wildlife Commission, notes that a University of Florida research team recorded one hundred active kite nests—a record number—in the STAs during 2014. The number of nests varies from one year to the next. Based on data through 2016, Garrett estimates that a total of 140 successful nests have raised chicks in the STAs since the kites were first found nesting there in 2010.[18] That's a significant contribution to the survival of a bird whose total US population is estimated at only 1,200 individuals.

Though water levels shift—deeper in the summer rainy season, shallower in the dry of winter—the STAs are kept flooded year round to maintain the plants that work to cleanse the water. Permanent water is good for apple snail populations, and for snail kites too. This is far different from the undisturbed Everglades, however, much of which dried out for months at a time before the drastic re-plumbing of south Florida. Garrett and other managers work to control water levels in the STAs during the months when black-necked stilts are breeding. These elegant wading birds, which stalk through the water on slender, bright-red legs, nest on the surface of muddy pond banks. An uncontrolled flood could wipe out hundreds of stilt chicks.

The threatened wood stork is also adapted to the ancient rise and fall of water in the undisturbed Everglades. A stork feeds by probing with its bill in murky water: Its jaws snap shut with lightning speed when the bird senses a fish. The

Figure 11.4 Apple snail shells from a Stormwater Treatment Area. Photo by Sharon Levy.

species traditionally feasted in small pools where fish concentrate when wetlands dry out in winter, making the stork's hunt-by-feel tactic more effective. The massive re-engineering of south Florida's water flows has hit wood storks hard, eliminating the seasonal changes that shaped the birds' survival strategy. We saw several storks on our tour of STA 5/6, exploiting the abundant small fish there. Still, the long-term survival of the US wood stork population will likely require major changes in the plumbing humans have imposed on south Florida, to allow a rebirth of seasonally dry wetlands—a habitat the STAs cannot replace.

Every September, thousands of sandpipers and stints arrive at the edge of Port Phillip Bay, near Melbourne, Australia. They gather in numbers seen nowhere else on the Australian coast. Exhausted from their long migration—many of these tiny shorebirds breed as far away as northeastern Siberia or western Alaska—they settle in for some serious eating. At low tide, they spread out on the mudflats at the bay's edge, probing into the muck with their sensitive bills. The bay mud is a rich source of small invertebrates, made even richer by discharges from the Western Treatment Plant (WTP), which for decades has handled the sewage generated by millions of people living in Melbourne. Long-term discharge of treated sewage has given the bay's nearshore ecosystem a nutritional jolt, enhancing the growth of algae and the small creatures that feed on it, creating a feast for hungry shorebirds.

The WTP has a distinction unique among wastewater facilities: It's recognized as the centerpiece of a conservation site of international importance under the Ramsar Convention on Wetlands due to its intense use by water birds. In addition to helping feed large numbers of native Australian shorebirds and migrants from the far north, the plant contains a series of sewage treatment lagoons, some of them built as long ago as the 1930s, which have become critical habitat for waterfowl.

Unlike the STAs in south Florida, lagoons at the WTP lack emergent plants like cattail. Called waste stabilization ponds, these open-water systems are seen by some experts as the simplest, least energy-intensive means of treating sewage, and are widely used in the developing world.[19] In the lagoons, bacteria break down organic compounds, while algae and zooplankton flourish on the nutrients in the sewage. The WTP is the world's largest array of wastewater stabilization ponds; the second largest is in Nairobi, Kenya.

Many Australian ducks and geese breed at ephemeral wetlands in the continent's interior, which dry up at the end of the wet season. In the dry months, the birds head to the coast, seeking refuge and food in permanent wetlands. More than a third of southeast Australia's original coastal wetlands have been lost to drainage and development. The treatment lagoons at the WTP are reliably flooded with high-nutrient water, booming with algae and aquatic invertebrates. High nutrient loads in most of the ponds have kept them empty of fish, leaving the bonanza of food available to birds.

Recent studies by Christopher Murray and Andrew Hamilton of the University of Melbourne have found that wastewater treatment ponds, at the WTP and elsewhere in southeast Australia, host more abundant and diverse bird populations than remaining natural wetlands.[20] "Over a period of 22 years, we found significantly more species, and higher numbers of birds, in wastewater ponds," explains Hamilton. Like south Florida's STAs, the treatment ponds attract birds because of their intense biological productivity, fed by the high nutrient levels in polluted water.

In a single day of observation at Pond 9, a 109-hectare segment of the Lake Borrie lagoon system at the WTP, Hamilton has counted as many as five thousand pink-eared ducks,[21] striking birds that sport zebra-striped plumage and hold their long, lobed bills just under the surface to filter plankton out of the water. They swim alongside large flocks of teal, shovelers, grebes, and swans; Hamilton has recorded up to twenty thousand birds at a time on Pond 9. Perhaps the most dramatic example of a species that relies on habitat at the WTP is the blue-billed duck, a diving bird that forages for invertebrates on pond bottoms. In 2002, when the world population of this duck was estimated to be about twelve thousand individuals, more than thirteen thousand were counted at the WTP. The plant has also hosted twenty-five thousand hoary-headed grebes at a time, dwarfing recorded numbers at any other Australian wetland.[22]

This mind-boggling abundance does not represent normal conditions in a natural wetland. "Waste stabilization ponds support a very different type of community when it comes to the balance between numbers and diversity of various species," notes Hamilton. "They're great for water birds, but not for amphibians and arthropods, which need emergent vegetation . . . Because we've lost so much natural habitat, we're happy to have a place like the Western Treatment Plant, a sort of a McDonald's for water birds."

These treatment wetlands, and the birds that use them, exist on the knife edge between artificial abundance and the deadly effects of eutrophication. In the series of interconnected ponds used in waste stabilization systems, the first one or two ponds will be anaerobic, the free oxygen drained from the water by the decomposition of heavy loads of organic matter. In these extreme conditions, anaerobic bacteria digest pollutants, but few aquatic creatures survive. Farther along in the sequence, where the water is cleaner, aerobic ponds support abundant food for water birds. Murray and Hamilton have documented the highest abundance and species richness of water birds in the aerobic ponds at the cleaner end of treatment systems, both at the WTP and in the Goulborn Valley, a major agricultural area in Victoria.[23]

Port Phillip Bay remains surprisingly healthy despite long-term discharges from the WTP and other sources in sprawling metropolitan Melbourne. An intensive study of the bay's ecology[24] explains this happy accident. The bay is shallow, and its waters remain clear. Sunlight penetrating to the bay floor allows microalgae to flourish on the bottom. Thick mats of benthic microalgae capture nutrients before they can escape back into the water column to fuel algal growth there, which would cloud the water.

Meanwhile, the bottom-growing algal mats feed a diverse community of invertebrates—several hundred known species—that live in the sediment. These small creatures burrow through the sediment, mixing organic forms of nitrogen into the muck. The nitrogen that isn't held in the sediment is transformed to N_2 gas by bacteria and diffuses up through the water and into the atmosphere.

A model of nutrient cycling in the bay created by scientists with the Commonwealth Scientific and Industrial Research Organization, Australia's national science agency, predicted that continued nutrient loading could push Port Phillip Bay over the brink, clouding the water, smothering the benthic algal mats, and killing many species of benthic invertebrates. To prevent this, the WTP was ordered to improve its wastewater treatment, to decrease the amount of nitrogen it released to the bay.

In 2003–2005, activated sludge systems were installed in two of the WTP's newer treatment lagoons. Some of the older lagoons, including the Lake Borrie ponds, were removed from the sewage treatment process and began to receive fully treated effluent. Ecologists feared that cleaner water would mean less algae and invertebrates for waterfowl to dine on, reducing Lake Borrie's value as bird habitat. While one Australian law mandated the upgrade to the sewage treatment process to protect Port Phillip Bay, another required that plant managers track and mitigate any resulting impacts on bird populations in the WTP.

To compensate for a possible decline in abundance of shorebird prey in Port Phillip Bay's tidal mudflats, managers at the WTP created shallow "conservation ponds" that provide permanent places for shorebirds to roost and feed, unaffected by shifting tides. Early studies had suggested that filter-feeding birds like the pink-eared duck, which seemed to prefer high-nitrogen waters, might be most affected by the sewage treatment upgrade. Surveys of waterfowl in the first years following installation of the activated sludge plants showed an overall decline in the time birds spent feeding on Pond 9 of the Lake Borrie system—a trend that was most pronounced among bottom-feeding species like chestnut teal, gray teal and Australian shelduck. Birds at the WTP also changed their habits, spending more time on ponds downstream of the new activated sludge plants that had once been too eutrophic to attract them. The numbers of birds in the Lake Borrie system dropped. Managers responded by trying various ways of bringing more nutrients back into Lake Borrie, and ultimately decided to build a new pipeline that will return sewage flows there.[25]

The major changes at the WTP coincided with an epic drought that struck eastern Australia from 1997 to 2009.[26] Many natural wetlands went dry; at the height of the drought, in 2008, aerial surveys found that about 70 percent of all waterfowl in the state of Victoria were at the WTP, sheltering on its reliably flooded ponds.[27]

More recent data show that climate—specifically the drought and its end, which saw flooding of long-dry natural wetlands in the continent's interior—drove many of the variations in water bird use of habitat at the WTP. "In hindsight, having seen how birds rebounded after the drought, I reckon we'd have great difficulty arguing for the major expense of the Lake Borrie sewage pipeline on the basis of bird habitat alone," notes William Steele, senior biodiversity scientist for Melbourne Water, the agency that manages the WTP. The pipeline will likely end up paying its way, as continuing human population growth in Melbourne creates a need for more treatment lagoons. It's a point lesson in the complexities of managing treatment wetlands and the birds that flock to them.

In the US, the federal government has had a policy of no net loss of wetlands since 1989. Under regulations enforced by the US Army Corps of Engineers, those who destroy wetlands are required to create new ones in mitigation. On paper, constructed wetlands have begun to balance out the acreage of habitat lost to ongoing development. Yet research over the last fifteen years has made clear that for wildlife, wetland habitats constructed under the "no net loss" policy often don't replace those lost. The needs of amphibians have often been overlooked, says Christopher Shulse, an ecologist with the Missouri Department of Transportation who spends much of his time monitoring wildlife in artificial wetlands. "Amphibians," he notes, "are the canary in the coal mine."

Wetlands are where frogs and salamanders mate and lay their eggs, where tadpoles grow up and sprout the limbs that allow them to emerge onto land. Many species also need healthy upland habitat—forest or grassland—to survive as adults. So the health of amphibian populations reflects not just the state of a single pond, but also that of the surrounding landscape.

Constructed wetlands in Missouri make excellent habitat for one or two hardy species: Bullfrog and green frog do well in steep-banked, permanent ponds, alive with fish that prey on frog eggs and tadpoles. The widespread creation of farm ponds has extended the range of the bullfrog into hillsides where it was not traditionally found. But for many native amphibians, such ponds are a dead end. "The wetland types credited with reversing the national trend toward wetland loss most often have steep banks, permanent water, and little vegetative cover," says Shulse. "They may harbor plenty of stocked fish and bullfrogs, but most native amphibians can't survive in these habitats."

If they are to work for amphibian conservation, artificial wetlands must become more diverse. Native frogs and salamanders need a range of wetlands, from the ephemeral to the permanent, scattered liberally across a landscape that includes upland habitats the creatures can use in adulthood.

In many urban areas, it has become nearly impossible to find a natural wetland, and the only homes available to amphibians are stormwater retention ponds, built to filter toxins from the polluted water that runs off city streets. Joel Snodgrass, head of the wildlife department at Virginia Tech, has spent years studying amphibians in these artificial habitats. In a study of forest and suburban wetlands in a rapidly developing area of Baltimore County, Maryland, Snodgrass and his colleague Adrianne Brand found that most amphibian breeding activity took place in manmade wetlands, and that these were the only places where tadpoles survived to metamorphosis.[28] Survival of eggs and larvae was highest in ponds that went dry for part of the year.

Keeping native amphibians alive in an urbanizing landscape will mean managing stormwater ponds for their benefit. Just how to do this is a difficult question, since stormwater ponds collect a stew of poisons: metals, nutrients, and toxic hydrocarbons left behind by the steady communal drip of gasoline, antifreeze, fertilizers, and pesticides. One ingredient is zinc, which wears out of automobile tires: An estimated ten thousand tons of zinc was released to US roadways through tire wear in 1999. Snodgrass's group found that wood frog (*Rana sylvatica*) larvae exposed to zinc from tire debris showed decreased hatching success, slowed development, and lower weight at metamorphosis, which may mean their chances of adult survival are reduced.[29]

American toads (*Bufo americanus*; Fig. 11.5) thrive in suburban Baltimore, but the wood frog is vanishing. Part of the problem is that the forest habitat where wood frogs spend their adulthood is dwindling. In his lab, Snodgrass found that habitat loss is not the sole threat. He exposed larval wood frogs and American toads to sediments from Baltimore-area stormwater ponds, contaminated with chromium, copper, nickel, zinc, and road salts. The toads raised in these toxic sediments suffered some impacts: They were smaller when

Figure 11.5 An American toad, one of the few amphibian species thriving in suburban Baltimore, Maryland. Photo by Joel Snodgrass.

they metamorphosed into adults than their counterparts raised on clean sand. None of the wood frog larvae exposed to the toxic sediments survived to metamorphosis at all.[30]

Recent evidence suggests that salt, used to melt ice off roads in winter, is the deadliest toxin for sensitive species like the wood frog. The concentration of road salt in stormwater ponds appears to be a major factor determining which amphibians can successfully reproduce. Stormwater ponds are often located next to remnant forest habitats, which might help the wood frog to hang on if its young weren't dying of salt exposure.[31]

Researchers in Edmonton, Canada, also found evidence that stormwater ponds act as ecological traps. In 2009, they recorded that larval wood frogs survived to adulthood in 100 percent of the natural wetlands surveyed. In Edmonton's stormwater ponds, the figure dropped to 32 percent.[32]

Andrew Hamer, an ecologist at the Australian Research Centre for Urban Ecology, has spent many nights listening at the edges of stormwater treatment ponds in Melbourne, identifying frogs by their distinctive courting songs. In a study relating habitat structure in these manmade ponds to frog survival,[33] his results echoed Shulse's concern that one type of habitat cannot sustain all amphibians. Some species, like the spotted grass frog, mate only in wetlands thick with emergent plants. By contrast, the southern bullfrog makes its love call, a resonant *bonk* like a note struck too hard on a banjo, while floating on submerged plants in ponds with steep, barren banks.

Most of Melbourne's stormwater ponds have been designed to hold water permanently, and they harbor thriving populations of *Gambusia holbrooki*, known on its native turf as the mosquitofish. Introduced from the southeastern US in

1925, *G. holbrooki* has become an environmental disaster in Australia, where it is now called the plague minnow. (Carried around the globe in the false hope of mosquito control, *G. holbrooki* is the most widespread fish on the planet, listed on the International Union for Conservation of Nature's list of one hundred worst invasive species.) The little fish flourishes in polluted water, feeding voraciously on native aquatic creatures.

Though he has heard other native frogs calling in the stormwater ponds, Hamer found that only tree frog, striped marsh frog, and southern bullfrog larvae survived to adulthood in the presence of plague minnows.[34] In Melbourne, as in Maryland and Edmonton, stormwater ponds can act as ecological traps.

The growing body of evidence on amphibian life and death should ultimately lead to changes in stormwater pond management. Melbourne's ponds could be made more amphibian-friendly by altering some so that they dry out for part of the year, preventing the establishment of plague minnows. That kind of habitat diversity has already been accomplished in the Baltimore area, but high contaminant levels remain a threat. Where the goal is solely to treat water pollution, toxins will accumulate at levels that form ecological traps even for the hardiest toad. In those places, barriers might be needed to keep amphibians out.

Finding ways to make stormwater ponds serve as healthy frog habitat will be a complex mission, moving beyond the system's original design as a way of simply slowing polluted water and allowing contaminants to settle out. "We'll have to figure out where the critical thresholds are for pollutant levels, and for combinations of pollutants: metals, pesticides, fertilizers, salts, polyaromatic hydrocarbons," says Snodgrass. "It's a soup of pollutants that's difficult to understand."

Oil sands mining in Alberta, Canada, is producing vast amounts of an even more toxic stew. Separating bitumen from oil sand ore takes 2 to 2.5 cubic meters of fresh water for every cubic meter of synthetic crude oil produced.[35] As a result, large holding ponds full of liquid tailings are accumulating. These oil sands process-affected waters (OSPW) are heavily contaminated with napthenic acids, polycyclic aromatic hydrocarbons, metals, and salt.

Constructed wetlands are part of the plan to restore the blasted landscape left behind by oil sands mining. Thus far a few wetlands have been deliberately built, while others have formed spontaneously. Research on these wetlands reveals both the remarkable resilience and purifying abilities of native plants and the ecological complexities of creating wildlife habitat with heavily contaminated water.

Native sedges can thrive even in the toxic soup of OSPW.[36] Judit Smits, a wildlife toxicologist at the University of Calgary, has tracked the impacts of life in these reclaimed wetlands on birds and frogs. In newly formed oil sands wetlands, wood frog tadpoles suffer delayed metamorphosis—a problem that's linked to effects on their thyroid glands, and that reduces their odds of surviving to successfully reproduce.[37] Their blood contains high levels of cytochrome P450 enzymes, proteins produced by the liver in response to toxin exposure—a sign that precious energy is being diverted from normal growth and development to the metabolism of poisons.

Over time, sunlight and microbial metabolism degrade the organic pollutants, while metals and other toxins sink to the bottom of wetland ponds and are held in the sediment. Seven years in, the worst of the toxic impacts on wildlife fade. Wood frog tadpoles raised in OSPW wetlands more than seven years old do as well as those raised in undisturbed reference habitats.

Smits has studied tree swallows living in nest boxes among reclaimed OSPW wetlands. The birds feed on aquatic insects that spend their larval phase in the wetland itself. Swallows living in new wetlands appear hard-hit by the toxins in their diet. Like wood frog tadpoles, tree swallow nestlings on newly reclaimed wetlands grow slower than those in reference sites and show higher levels of cytochrome P450 enzymes in their blood. In the spring of 2003, when harsh weather triggered a widespread die-off of nestlings, the odds of survival on the most heavily polluted sites was ten times lower than at the control site. Tree swallows on OSPW wetlands are also heavily infested with blow flies; nestlings carried parasitic burdens twice as high as those recorded on the reference site. Some of the toxins in OSPW can impair the birds' immune response to parasites. "On the reclaimed sites," says Smits, "the birds have to use energy to cope with toxicants. So they don't have the resilience or energy reserves of birds living in natural wetlands."

Smits believes constructed wetlands can be a viable way of rehabilitating the post-mining landscape in Alberta. The caveat is that OSPW wetlands will be highly toxic for the first seven years of their existence. "If we want to protect the animals, we have to avoid them getting onto the younger wetlands," she says. "Those areas would need to be fenced off for a few years." So far, there's no requirement that reclamation keep pace with oil extraction, and vast quantities of liquid mine tailings are piling up. Aside from a few wetlands built for research and aged enough to have lost their toxicity, Alberta's oil sands mine areas (1,670 square kilometers were actively being mined or approved for development as of 2013)[38] promise to remain a black hole for wildlife for years to come.

That bleak scene is hard to imagine while walking the trails in the Arcata Marsh. The city's constructed wetlands remain a hotspot for wildlife, hosting large populations of migrating ducks and shorebirds in spring and fall. Mallards, green-winged and cinnamon teal, and Canada geese breed successfully in the summer, and families of river otters play here year-round. But the future of Arcata's popular wetland is now in doubt.

NOTES

1. Baden, S., Andreas Emanualsson, Leif Pihl, Carl-Johan Svensson, Per Aberg (2012). "Shift in seagrass food web structure over decades is linked to overfishing." *Marine Ecology Progress Series* **451**: 61–73.
2. Hughes, B. B., Ron Eby, Eric Van Dyke, et al. (2013). "Recovery of a top predator mediates negative eutrophic effects on seagrass." *PNAS* **110**(38): 15313–15318.
3. Baden, S., Andreas Emanualsson, Leif Pihl, Carl-Johan Svensson, Per Aberg (2012). "Shift in seagrass food web structure over decades is linked to overfishing." *Marine Ecology Progress Series* **451**: 61–73.
4. Moksnes, P.-O., Martin Gullstrom, Kentaroo Tryman, Susanne Baden (2008). "Trophic cascades in a temperate seagrass community." *Oikos* **117**: 763–777.

5. Baden, S., Christoffer Bostrom, Stefan Tobiasson, Heidi Arponen, Per-Olav Moksnes (2010). "Relative importance of trophic interactions and nutrient enrichment in seagrass ecosystems: a broad-scale field experiment in the Baltic-Skagerrak area." *Limnology and Oceanography* **55**(3): 1435–1448.
6. Reynolds, P., J. Paul Richardson, J. Emmett Duffy (2014). "Field experimental evidence that grazers mediate transition between microalgal and seagrass dominance." *Limnology and Oceanography* **59**: 1053–1064.
7. Bertness, M. D., Brian R. Silliman (2008). "Consumer control of salt marshes driven by human disturbance." *Conservation Biology* **22**(3): 618–623.
8. Jefferies, R., Robert Rockwell (2002). "Foraging geese, vegetation loss and soil degradation in an Arctic salt marsh." *Applied Vegetation Science* **5**: 7–16.
9. Bertness, M. D., Brian R. Silliman (2008). "Consumer control of salt marshes driven by human disturbance." *Conservation Biology* **22**(3): 618–623.
10. Ibid.
11. Piehler, M.F., A.R. Smyth (2011). "Habitat-specific distinctions in estuarine denitrification affect both ecosystem function and services." *Ecosphere* **2**(1): 1–13.
12. O'Meara, T., Suzanne Thompson, Michael Piehler (2015). "Effects of shoreline hardening on nitrogen processing in estuarine marshes of the US mid-Atlantic coast." *Wetlands Ecology and Management* **23**: 385–394.
13. Hughes, B.B., Mattew D. Levey, Monique C. Fountain, et al. (2015). "Climate mediates hypoxic stress on fish diversity and nursery function at the land-sea interface." *PNAS* **112**(26): 8025–8030.
14. Beck, T., Dale Gawlik, Elise Pearlstine (2013). "Community patterns in treatment wetlands, natural wetlands, and croplands in Florida." *Wilson Journal of Ornithology* **125**(2): 329–341.
15. Izuno, F.T. (1987). "A brief history of water management in the Everglades Agricultural Area." Institute of Food and Agricultural Sciences, University of Florida, **Circular 815** (http://share.disl.org/stanton/Shared%20Documents/Everglades/Izuno_Water%20management%20in%20the%20EAA.pdf).
16. Mitsch, W., Li Zhang, Darryl Marois, Keunyea Song (2014). "Protecting the Florida Everglades wetlands with wetlands: can stormwater phosphorus be reduced to oligotrophic conditions?" *Ecological Engineering* 80: 8–19.
17. Entry, J.A., A. Gottlieb (2014). "The impact of stormwater treatment areas and agricultural best management practices on water quality in the Everglades Protection Area." *Environmental Monitoring and Assessment* **186**: 1023–1037.
18. Garrett, B. (2017). "Appendix 5B-3: Summary of Stormwater Treatment Area Black-necked Stilts and Other Protected Birds during the 2016 Nesting Season." 2017 South Florida Environmental Report, Volume I (http://apps.sfwmd.gov/sfwmd/SFER/2017_sfer_final/v1/appendices/v1_app5b-3.pdf).
19. Mara, D. (2003). "Domestic wastewater treatment in developing countries." London and Sterling, Virginia: Earthscan.
20. Murray, C., Richard Loyn, et al. (2012). "What can a database compiled over 22 years tell us about the use of different types of wetlands by waterfowl in south-eastern Australian summers?" *Emu* **112**: 209–217.
21. Hamilton, A., Iain Taylor, Graham Hepworth (2002). "Activity budgets of waterfowl (Anatidae) on a waste-stabilisation pond." *Emu* **102**: 171–179.
22. Richard Loyn, personal communication, 2015.

23. Murray, C., Sabine Kasel, Erin Szantyr, Regan Barratt, Andrew Hamilton (2014). "Waterbird use of different treatment stages in waste-stabilisation pond systems." *Emu* **114**: 30–40.
24. Harris, G., G. Batley, D. Fox, D. Hall, P. Jernakoff, et al. (1996). "Port Phillip Bay Environmental Study." Collingwood, Victoria: CSIRO Publishing.
25. Steele, W., S. Harrow (2014). "Overview of adaptive management for multiple biodiversity values at the Western Treatment Plant, Werribee, leading to a pilot nutrient addition study." *Victorian Naturalist* **131**(4): 128–146.
26. Loyn, R.H., D.I. Rogers, R.J. Swindley, K. Stamation, P. Macak, P. Menkhorst (2014). "Waterbird monitoring at the Western Treatment Plant, 2000–12: The effects of climate and sewage treatment processes on waterbird populations." *Arthur Tylah Institute for Environmental Research Technical Report Series* **256**.
27. Ibid.
28. Brand, A., Joel Snodgrass (2010). "Value of artificial habitats for amphibian reproduction in altered landscapes." *Conservation Biology* **24**: 295–301.
29. Camponelli, K., Ryan E. Casey, Joel W. Snodgrass, Steven M. Lev, Edward R. Landa (2009). "Impacts of weathered tire debris on the development of *Rana sylvatica* larvae." *Chemosphere* **74**: 717–722.
30. Snodgrass, J., Ryan E. Casey, Debra Joseph, Judith A. Simon (2008). "Microcosm investigations of stormwater pond sediment toxicity to embryonic and larval amphibians: variation in sensitivity among species." *Environmental Pollution* **154**: 291–297.
31. Gallagher, M., Joel Snodgrass, Adrianne Brand, Ryan Casey, Steven Lev, Robin Van Meter (2014). "The role of pollutant accumulation in determining the use of stormwater ponds by amphibians." *Wetlands Ecology and Management* **22**: 551–564.
32. Scheffers, B., Cynthia Paszkowski (2013). "Amphibian use of urban stormwater wetlands: the role of natural habitat features." *Landscape and Urban Planning* **113**: 139–149.
33. Hamer, A., Phoebe J. Smith, Mark J. McDonnell (2012). "The importance of habitat design and aquatic connectivity in amphibian use of urban stormwater retention ponds." *Urban Ecosystems* **15**: 451–471.
34. Hamer, A., Kirsten Parris (2013). "Predation modifies larval amphibian communities in urban wetlands." *Wetlands* **33**(4): 641–652.
35. Toor, N., Eric Franz, Phillip Fedorak, Michael MacKinnon, Karsten Liber (2013). "Degradation and aquatic toxicity of napthenic acids in oil sands process-affected waters using simulated wetlands." *Chemosphere* **90**: 449–458.
36. Raab, D., Suzanne Bayley (2013). "A *Carex* species-dominated marsh community represents the best short-term target for reclaiming wet meadow habitat following oil sands mining in Alberta, Canada." *Ecological Engineering* **54**: 97–106.
37. Hersikorn, B., Judit E.G. Smits (2011). "Compromised metamorphosis and thyroid hormone changes in wood frogs (*Lithobates sylvaticus*) raised on reclaimed wetlands on the Athabasca oil sands." *Environmental Pollution* **59**: 596–601.
38. Raab, D., Suzanne Bayley (2013). "A *Carex* species-dominated marsh community represents the best short-term target for reclaiming wet meadow habitat following oil sands mining in Alberta, Canada." *Ecological Engineering* **54**: 97–106.

12

Of Time and the Wetland

At the oldest of Arcata's treatment wetlands, it's now possible to walk on water. Over three decades of filtering sewage, Arcata's wetland cells have developed floating mats of dead cattail stems and leaves underlain by living roots, resilient enough to support a person's weight. The short journey across Treatment Wetland 3 is a strange experience, like walking on a soggy trampoline. Water seeps through the cattail mat and into footprints. On a February day, a dense maze of brown cattail stems stretches twelve feet above the wetland's surface, their shaggy brown seedheads waving in the breeze (Fig. 12.1).

A stroll across the treatment wetland is as close as a modern American can hope to get to the feel of the floating tule islands that William Finley camped on in the upper Klamath Basin in 1905, and that crowded California's unspoiled marshes before the Gold Rush. The floating mats in Arcata were created by accident when the city's treatment plant operators increased the depth of the treatment marshes, part of an effort to improve their declining performance. To their surprise, the dense growth of cattail rose off the bottom and continued to thrive, roots dangling in the water. The wetlands have aged.

"Arcata's is the grandmother municipal treatment wetland," says David Austin, an environmental engineer with CH2M Hill who specializes in treatment wetlands design. Austin remembers studying the Arcata wetlands as a student at University of California at Davis in the 1990s. "It was a pioneering system. Now it's an old design—one that wouldn't be used today."

In 2016, three decades after Bob Gearheart's unconventional marshes began cleaning Arcata's sewage, the city's wastewater plant faced a crisis. During the cold rains of winter, the system often failed to perform to the standards set in its discharge permit. Every part of the plant had aged to the point where its performance was in decline. At the headworks, the two giant Archimedes screws that push raw sewage uphill through a coarse screen had been running for decades; their metal housings were rusting away. The concrete basins of the primary clarifiers—where suspended solids settle out of raw sewage—were cracked, their bottoms eroded by long years of use.

Repairing or replacing these mechanical parts of the system may be costly, but it's a straightforward process. Refurbishing the natural parts of the treatment

Figure 12.1 View from the floating cattail mat on Treatment Wetland 3. Photo by Sharon Levy.

plant, the oxidation ponds and treatment wetlands, is vital, but the path forward is less clear. The bottoms of the treatment wetlands were covered in decades' worth of a nutrient-rich soup of decaying algae and plant parts, which impaired their ability to filter suspended solids and nutrients from wastewater.

All sewage treatment plants age. Mechanical systems, like the activated sludge plants common in big cities, need regular repairs and often require major upgrades after thirty years or so. One of the daunting problems facing Arcata is that nobody knows just how to handle the millions of gallons of organic sludge and the thousands of pounds of overgrown cattail in the treatment wetlands.

The stakes are high. Water quality rules have become more stringent in the decades since the treatment wetlands started up, and every few years Arcata gets a bill from the North Coast Regional Water Quality Control Board for hundreds of thousands of dollars in fines triggered by occasional violations of the city's discharge permit.

"The wastewater treatment plant is our most critical piece of infrastructure," says Mark Andre, Arcata's director of environmental services. The city's future growth depends on the plant working reliably: Every new sewer hookup, or a shift from a single-family home to an apartment building, increases the volume of sewage. Should the city continue to violate its permit, the regional board has the power to impose a building moratorium, as it did during the battle over the regional sewage system in the 1970s.

Rebuilding or refurbishing a sewage plant is one of the biggest investments any US community will make—and forty-five years after the Clean Water Act (CWA) was passed, many cities in California are struggling to find the funds. Rebuilding Arcata's sewage treatment system is expected to cost somewhere in the neighborhood of $30 million to $40 million. In cities that have conventional treatment systems, upgrades are costly but the technical aspects are clear. Arcata's system is unique, and bringing it up to modern standards poses some unprecedented challenges (Fig. 12.2).

In spring and summer the treatment system is at its best. In the oxidation ponds, algae flourish on the rich supply of nutrients. As they photosynthesize, the algae release oxygen into the water, enabling aerobic bacteria to thrive as they break down organic compounds. In engineering lingo, the ponds reduce biochemical oxygen demand (BOD)—the main goal of secondary treatment. When wastewater flows on to the treatment wetlands, dense leaves of cattail shade out the algae, which die and settle to the bottom, dramatically lowering the concentration of suspended solids. In the warm season, when little rain falls, the effluent flowing out of the natural treatment system is often cleaner than is required under the city's discharge permit.

As the days become short and cool, however, algal growth declines and dissolved oxygen levels fall. In the wetlands, the cattails turn brown and stop growing. The rains come, dumping cold water into the system, so that the populations of microbes that do the work of decomposing and transforming pollutants dwindle and their metabolism slows. Growth yields to decay, and the ponds and wetlands begin to release their internal load of nutrients from rotting plant matter.

At the same time, winter rains dramatically increase flows. Many of Arcata's sewer lines are old and cracked, so that rainwater seeps in. A significant number of homes in town have their gutter pipes routed to empty into the sewers. Added to this is the direct loading of rain onto the city's ninety acres of treatment ponds and wetlands. In winter, the plant receives up to 6 million gallons per day of sewage mixed with rain, all of which must be treated—but the treatment wetlands can handle only about 3 million gallons per day.

Most of the permit violations at the Arcata plant result from overloads of rainwater that force operators to bypass the treatment wetlands, releasing effluent that exceeds permitted levels of suspended solids and of BOD. Some of the violations come from the chlorine disinfection system, meant to kill any pathogens that survive passage through the biological sieve of oxidation ponds and treatment wetlands. Gearheart, the wetlands' founding father, has been opposed to chlorination for decades, because in addition to sucking up energy and presenting a hazard to workers, the process produces toxic chlorinated hydrocarbons. Some of the plant's recent violations have been for release of one such poison, dichlorobromomethane.[1]

The extent of the treatment system's troubles became clear as the city worked to move past longstanding conflicts with the regional board. In the days of Arcata's sewage rebellion, the city built its series of three large marshes in defiance of the state water board's directives. Using a grant obtained from the California Coastal Conservancy, Arcata built the wetlands before the board-approved wetlands

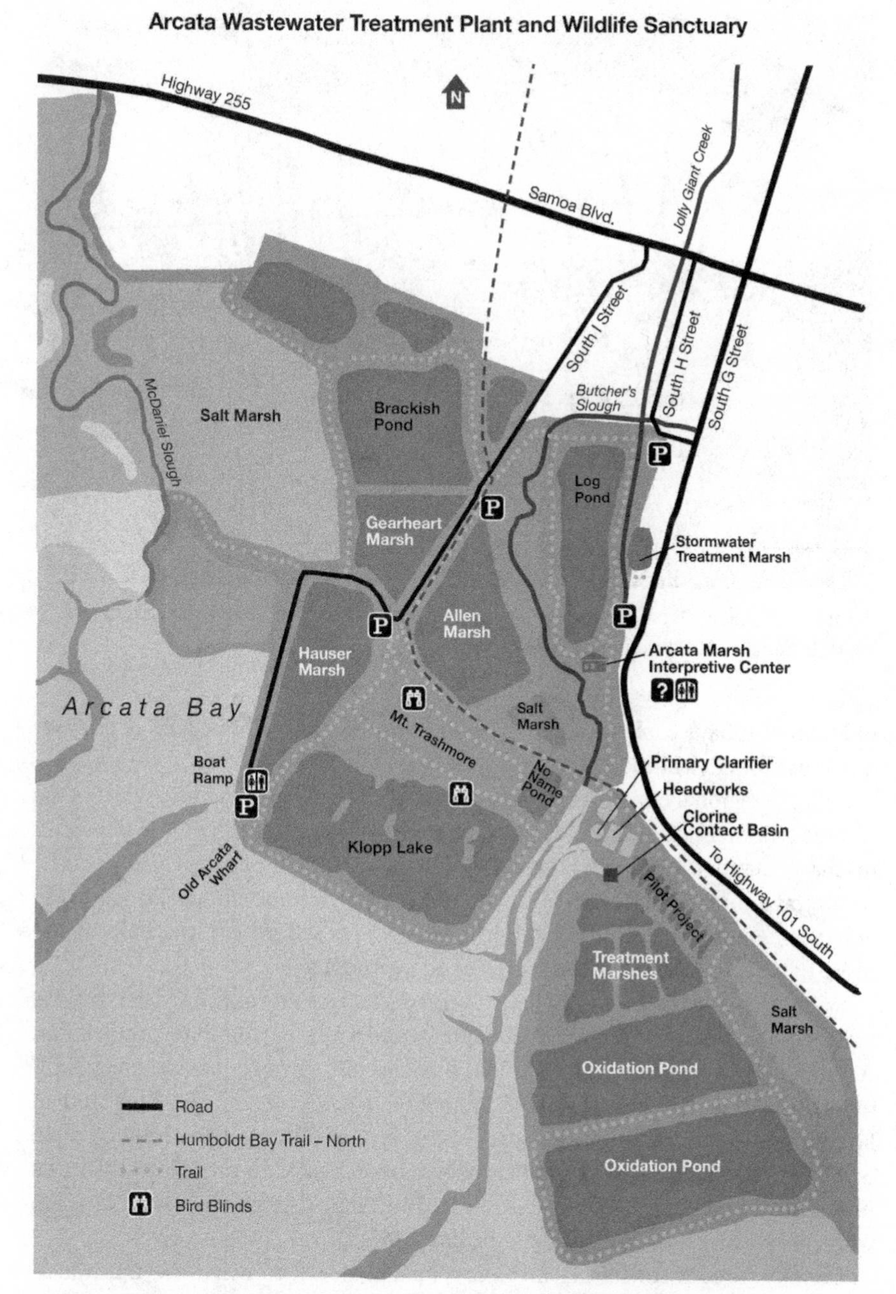

Figure 12.2 Map of Arcata Wastewater Treatment Plant and Wildlife Sanctuary.

treatment pilot project had been constructed. The marshes, dubbed Allen, Gearheart and Hauser after the major players in Arcata's resistance, were watered for the first few years with flow from a nearby creek because the board had not approved their use to help treat wastewater (Fig. 12.3). In time, the board accepted these three marshes as part of the Arcata treatment system's official "enhancement"

Figure 12.3 Allen Marsh, one of the "enhancement" wetlands at the Arcata Marsh and Wildlife Sanctuary. Photo by Maya Scanlon.

of Humboldt Bay's ecology, based on their importance as wildlife habitat and frequent use for environmental education.

Still, the regional board saw these marshes as natural habitats, not as part of the sewage treatment system. Board staff insisted that the city's effluent be treated to discharge standards *before* it flowed through the three enhancement marshes. The result was a complex series of treatment paths in which some effluent flowed to the enhancement marshes and some didn't, and some of the effluent was chlorinated more than once, increasing the risk of toxic byproducts.

In 2012, the city at last negotiated a series of changes that staff had been hoping to achieve for decades. The regional board issued a new permit that specified that Arcata's entire sewage flow should travel a single path, through primary treatment, oxidation ponds, treatment wetlands, and enhancement wetlands, which were at last acknowledged as an integral part of the system. The chlorine contact basin would be retired, and instead effluent would be treated with ultraviolet light as a final disinfection step, before it was released to a brackish marsh created in 2008, where water surges in and out with the bay's tides.

The city hired a consulting firm, Carollo Engineers, to help with the transition to ultraviolet disinfection. The consultants analyzed the capacity of each part of the treatment system to remove BOD. They discovered that the chlorination process had been removing a significant proportion of BOD—in times of heavy loading, as much as 30 percent. Getting rid of the chlorine would mean a significant increase in the need for secondary treatment.

"We always knew that chlorination burns some BOD," explains Brad Finney, an environmental engineering professor at Humboldt State who's been studying the

Arcata treatment system with Gearheart since 1979. "The magnitude of that effect is surprising. It's not something you ordinarily bother to quantify, because you're hoping to meet BOD standards without that."

This new information came as a major setback. In addition to repairing or replacing the headworks and primary clarifier, Arcata had to find some way to significantly increase the plant's ability to remove BOD. The lion's share of BOD is taken up in the oxidation ponds—which are at the low point of their efficiency during the winter, when permit violations occur.

In the fall of 2015, Carollo presented its vision of how to bring Arcata's wastewater system into reliable compliance with discharge standards. Their solution was to build one or more oxidation ditches, a form of activated sludge treatment in which sewage moves through a series of raceways while large paddlewheels churn air into the liquid. Carollo has extensive experience with this technology, and it reliably controls BOD. It's also a concrete-and-steel, energy-intensive approach, the antithesis of the system Gearheart and his followers have built, studied, and expanded for more than thirty years. Many in Arcata feared the coming of activated sludge would mean the end of the city's cherished natural treatment system. In Carollo's preferred design, the oxidation ponds and wetlands would still exist , but would be reduced to a token status. They would continue to serve as wildlife habitat and a recreational site but no longer play a significant role in cleaning the city's wastewater.

What city staff had thought of as a few minor tweaks to the treatment plant would instead become a complete overhaul of the system. Soon the city was running out of money and time: In December 2016, the deadline for installing UV treatment came and went, while staffers tried to grapple with the unexpected magnitude of the problem. The plant continued to use chlorine for disinfection, while the city searched for a way forward.

Gearheart, now a professor emeritus at Humboldt State, acts as director of the Arcata Marsh Research Institute, headquartered in the old prefab metal building that once housed George Allen's aquaculture project. He remains actively involved in trying to find a way to keep the natural system going but declined to comment on the current state of affairs.

For Finney, who has worked closely with Gearheart on the marsh treatment system for decades, the Carollo proposal was backwards. "My personal opinion is that you start with fixing the system that you already have to make it run as well as possible," he says. "We're not using the existing ponds and wetlands to their full potential, and that can be improved. To instead put an oxidation ditch in as a giant Band-Aid doesn't make any sense."

Steve McHaney, now a project manager at the engineering firm GHD in Eureka, is a one-time Humboldt State engineering student who has collaborated with Gearheart on wetland treatment projects in Arizona, Palau, and Saipan. He feels he understands what the Carollo engineers are thinking—he worked for the firm earlier in his career. Wetlands are complex, and many details of their workings remain unknown. Because wetland treatment is unfamiliar to many engineers and remains something of a "black box" even to those who focus on it, it seems like a risky approach.

"It's not unreasonable to see that a more controlled process like activated sludge will make the plant's compliance more reliable," he notes. "But I'd hope that other options are fully explored before Arcata goes that route. Turning to activated sludge means the end of a dream." Gearheart and generations of his students have invested decades of effort to build an understanding of just how the wetlands do the work of treatment using only the energy provided by sun and wind. For them, resorting to conventional treatment would constitute a major defeat.

Greg Gearheart, the kid who grew up watching his father build Arcata's treatment wetlands, became an environmental engineer and has worked in California's water quality regulation system for decades. He's now in charge of data analysis for the State Water Resources Control Board, but he emphasizes that his comments reflect his own opinions, not his employer's.

"It's been an awkward fit from day one," he says, "to take a natural treatment system and make it work in the permit system."

In the early years of the CWA, the standard was that every city or large industry had to use the best available technology to treat its wastewater. In the 1970s, that best technology was assumed to be activated sludge treatment, which cultivates pollutant-digesting microbes from wastewater. Oxygen is pushed into sewage by churning it with high-energy-input agitators, allowing aerobic bacteria to do the work of decomposing organics and oxidizing nitrogen.

Water treatment requirements were eventually expressed in numbers: Dischargers were expected to provide effective secondary sewage treatment, defined as an 85 percent reduction in BOD and a thirty-day average of less than 30 mg per liter BOD and total suspended solids. These numbers are based on the performance of activated sludge systems. Engineers are taught to see activated sludge as the least risky, most reliable way to meet water quality standards.

Arcata's natural system demands far less energy, construction. and maintenance than an activated sludge plant, but its performance changes with the seasons. A few violations of permit standards have always been expected during the winter rains. That understanding was spelled out in the city's general plan, and it was acceptable because back in the 1980s and 1990s, occasional fines did not cost much. But that began to change in the late 1990s. The push for stricter enforcement of water quality laws began in southern California, just as it had a half-century before. The dead, gray waters of Dominguez Channel were again used to illustrate the problem (see Chapter 4). The Equilon oil refinery had routinely violated its National Pollutant Discharge Elimination System permit, releasing oil, grease, chlorine, and sulfides into the Channel. The Los Angeles regional board put off action for years. At last, in September 1999, it announced that Equilon would be fined $700,000, an amount negotiated with the refinery.[2] Environmentalists saw the penalty as too little, too late. The company, they alleged, had saved $2 million or more in treatment costs over the years, and $700,000 amounted to a slap on the wrist.

Statewide, the regional water quality control boards were failing to enforce the CWA. A 1997 study by the California Public Interest Research Group (CalPIRG) found that one in three large dischargers had violated their permits in a four-year period; the average violator stayed out of compliance for two years, often with its board's blessing.[3] Industries and sewage plants self-reported thousands of violations to the regional boards, but only a handful of fines were imposed—and many of these were reduced when dischargers complained. Sediments in the state's urban rivers and bays held dioxins, polychlorinated biphenyls (PCBs), arsenic, chromium, and other potent poisons. Bottom-dwelling fish in southern California bays were too toxic to eat.

"Due to a combination of weak tools, under-staffing and a lax attitude among some officials," concluded a CalPIRG report, "it is evident that enforcement of the Clean Water Act through fines or other penalties has never been a threat for many industries."

In 1999, the state legislature adopted the Clean Water Enforcement and Pollution Prevention Act, sponsored by San Francisco Assemblywoman Carole Migden, which created mandatory minimum penalties for violations of a discharge permit. The idea was based on a similar law enacted in New Jersey in 1990, which was followed by a more than 60 percent decrease in reported violations. The intent of the law was to force regulators to punish serious or chronic polluters. Greg Gearheart worked at the state board as a new bureaucratic machinery was built to enforce mandatory minimum penalties. In his view, the outcome has not been good. "A huge chunk of our resources go into writing and following up on mandatory minimum penalties," he says. "It's a mindless process that distracts the agency from focusing on the violators that are causing real pollution problems." Any violation now must be written up and fined—whether or not it created significant damage.

Gearheart questions whether mandatory penalties are fulfilling their intended mission. The law as it stands may deal a death blow to pond and wetland treatment systems like Arcata's. Traditional engineers have always sought the least risky, most predictable way of meeting discharge standards. Natural systems are dynamic, and seasonal changes can carry them outside the envelope of approved performance for removal of BOD, suspended solids, and ammonia—a form of nitrogen that is toxic to fish and invertebrates in fresh waters, which the North Coast regional board is just beginning to regulate. In the era of mandatory minimum penalties, consulting engineers are more and more averse to the risks that come with natural treatment. Consultants make more money when their clients go with conventional technologies that require construction of cement basins and steel aerators. "A natural system has a snowball's chance in hell of making it through the permitting process these days," says Gearheart, "because of these factors."

A regulatory standard is a blunt instrument, based on a mix of data and bureaucratic pressures. The available scientific literature often measures the toxicity of a particular pollutant in a single aquatic species under a certain set of circumstances, and that information has to be extrapolated to set discharge limits that apply to every treatment plant in a state. In the eyes of some of the people who know Humboldt Bay best, state standards don't make a great deal of sense here.

Chuck Swanson spent more than five years studying Arcata's wetland treatment system as a graduate student at Humboldt State. His thesis project examined nutrient cycles in Arcata Bay, the innermost arm of Humboldt Bay, and compared them to nutrient outflows from Arcata's effluent. His findings show that the contribution from Arcata's plant is dwarfed by the load of nitrogen, phosphorus, and silica carried into the bay from the ocean during spring and summer, when marine currents push nutrient-rich sediments up from the bottom in a process called upwelling. In fall and winter, when the efficiency of the city's natural treatment system declines and more nutrients are released, amounts of nitrogen coming from the wastewater plant are a fraction of those carried in stormwater running off the pastures and creeks around town. Most of the nitrogen in Arcata Bay in winter comes from the ecosystem's own internal load—the decay of eelgrass and phytoplankton that die back in winter in the bay just as cattail dies back in the treatment wetlands.[4]

"We have naturally high nutrient levels in the bay during spring and summer from upwelling in the ocean," says Swanson. "That's when Arcata Marsh is taking all the nutrients out of the water and contributing very little to the bay. In the winter, when the marsh releases more nutrients, Arcata Bay is flushed out because of higher tides and increased freshwater flows. The natural cycles of the wetland treatment system and those of Humboldt Bay are perfectly complementary."

In the end, though, Arcata must find a way to meet state standards year-round—even if those standards are based on much more polluted waters in much more urbanized parts of the state. "There are some who want to argue for exceptions to the discharge limits," says Andre. "That's not where I come from. We want to show strong environmental leadership—so it's awkward to ask for exceptions to the rules everyone has to follow."

Bob Gearheart has found a potential way to clean up and improve the performance of the treatment wetlands: a gizmo invented by a small company called Absolute Aeration, based in Greeley, Colorado. The company's Blue Frog circulator looks like a flying saucer as it floats on the water's surface (Fig. 12.4). Using a small motor and a minimal amount of electricity, the Blue Frogs push still water in a pattern that creates a stratified water column, with an anaerobic bottom layer. A population of anaerobic bacteria thrive in these conditions, digesting the organic gunk that has accumulated on the wetland floor. This approach is unorthodox—conventional systems rely on pushing oxygen into wastewater, an energy-intensive process. Following the discovery of anammox bacteria, which can break down organic compounds and pump nitrogen into the atmosphere in the absence of free oxygen, more engineers have begun to focus on anaerobic treatment systems.

In the spring of 2016, city staff had the dense vegetation cleared from the center of Treatment Wetland 3. Swanson mapped the topography of the sludge bed on the wetland's bottom and found its thickness varied from 1.3 to 2 feet. Two rented Blue Frog units began operating in the wetland in May. Three months later, in August, tests showed that the sludge layer in the wetland had shrunk considerably: About thirty-eight thousand cubic feet of gunk had vanished. The circulators also appeared to improve the wetland's ability to remove suspended solids and

Figure 12.4 Blue Frog circulators helping to reduce the load of organic sludge in Treatment Wetland 3. Photo by Sharon Levy.

BOD, both of which were significantly lower in effluent from Treatment Wetland 3 than in neighboring Treatment Wetland 1, which acted as a control.

The city continues to experiment with the Blue Frog circulators. Over time, they may prove to be an effective, low-cost way of clearing sludge from treatment wetlands, a process essential to restoring their efficient function. This could be an important step toward keeping the natural system going while meeting strict discharge standards.

Cleaned out and buffed up by Blue Frogs, Arcata's treatment wetlands should be able to handle more sewage. City staff also plan to convert the long-dormant aquaculture ponds where George Allen once raised his salmon smolts into an additional wetland cell, increasing the plant's treatment capacity. That would help, but would still fall short of the ability to handle the intense flows of a rainy winter.

For months, Arcatans proud of their treatment wetlands hung in suspense, waiting to see if their pioneering sewage system would endure. "Regulators," notes Scott Wallace, co-author of the 2009 textbook *Treatment Wetlands*, "try to apply a set of steady-state conditions to an ecosystem that's evolving over time." The result is the kind of conundrum facing Arcata.

George Tchobanoglous, an environmental engineer and professor emeritus at the University of California at Davis, reviewed Bob Gearheart's original proposal for a treatment wetland system in the 1970s, when he acted as a consultant for the state water board. Decades later, he's still promoting the radical notion behind

Arcata's wetland system—that wastewater is a valuable source of recyclable water, habitat, and energy.

"The science has advanced dramatically," he says. "You can use intensified wetland technology now and produce the same quality effluent you'd get with activated sludge, using a fraction of the energy."

Researchers have tried a number of ways of intensifying wetland treatment. In tidal flow wetlands, effluent floods the system and then is allowed to drain out. As a result, the wetland shifts between flooded, anaerobic conditions and a drained, aerobic state. This system prevents the buildup of sludge and supports the cycling of nitrogen from the water back to the atmosphere. "To my astonishment," says Austin, "after spending a few years developing the technology, I found that it had been worked out back in the 1890s, by an Ohio inventor named Cleophus Monjou. I've used some of his criteria to design flood and drain systems." The approach works well in many small communities, but likely cannot be scaled up to handle Arcata's peak flow of 6 million gallons per day.

Circulating effluent through a gravel bed that hosts a biofilm of aerobic bacteria can improve the performance of a constructed wetland. There are also a variety of wetland systems that have been given mechanical boosts of one sort or another. Wallace has designed subsurface flow wetlands in which air is pumped into the gravel bed through an irrigation hose, maximizing the action of aerobic bacteria that break down organic compounds. Austin is at work on a system that pumps pure oxygen into wetlands, achieving dissolved oxygen levels so high that the capacity to remove BOD is much increased.

These techniques are still emerging. In 2016, Tchobanoglous suggested that if Arcata was willing to take the kind of risks it did thirty years ago, the city's natural treatment system could survive. "There are real opportunities to intensify the design of the wetlands, and still keep the original function and approach," he said. But Arcata was already well behind schedule on changes the regional board expected, including the installation of UV disinfection as an alternative to chlorine. None of those closest to the problem—the city council, city staff, and Carollo Engineers—were interested in taking a major financial risk on an untried system. In explaining how they decided on recommending activated sludge treatment, Carollo employees wrote that "Any process not yet a proven technology with full scale installation experience was eliminated."[5]

At a public meeting in April 2017, Andre laid out options for upgrading Arcata's sewage plant. A series of aerators could be placed in one of the oxidation ponds, transforming it into a muddy facsimile of an oxidation ditch. This would create more secondary treatment capacity, but not enough to meet permit standards year-round. Carollo's facility plan recommended building two or three oxidation ditches, creating enough secondary treatment capacity to handle Arcata's sewage for the foreseeable future but rendering the treatment wetlands irrelevant.

The energy it takes to run this kind of system could more than triple the treatment plant's annual consumption of electricity. Andre called the energy projections "depressing," and said they represented "a quantum leap in energy consumption."

Susan Ornelas, a member of the city council, was Arcata's mayor until December 2017. She earned an engineering degree at Humboldt State in the 1990s, and studied in Bob Gearheart's classes. So she understands the complex problems facing the city's wastewater treatment system. She's also keenly aware of another coming crisis: Humboldt Bay is experiencing sea level rise at a rate higher than anywhere else in California.[6] Though no one can forecast the timing with precision, in the next few decades the bay is likely to rise up and drown the treatment plant entirely. Andre envisions a ring of fortifications built around the entire wastewater plant, including the treatment wetlands, to keep out the rising bay waters. That's an expensive proposition, which may or may not be approved by the Coastal Commission.

Ornelas had serious doubts. She ran a hand through her thick, graying hair as she explained the dilemma. "I have a deep love for Bob Gearheart," she said. "He wants to keep the wetlands going: they're his baby. We all love the marshes. But how can we invest $40 million in a place that's going to be underwater?"

She looked into the possibility of building any additional secondary treatment facility on an abandoned field inland from the bay's edge, about half a mile from the current oxidation ponds and treatment wetlands. City staff explored the idea, but found it expensive and impractical.

Arcata's wetlands were born out of a rebellion against wasteful conventional technology that would have consumed more energy than any other endeavor in Humboldt County. In the 1970s, when local activists fought the state's proposed regional sewer system, none of them knew about manmade global warming. Thirty-five years later, the evidence for human-induced climate change and rapid sea level rise is overwhelming. But the conventional engineers' response to Arcata's wastewater troubles is an energy-intensive activated sludge system, a choice that will cause the city's carbon footprint to balloon.

Conventional wastewater treatment systems use a lot of energy. A 2013 study found that they consume about 30 billion kilowatt-hours of electricity per year, or about 0.8 percent of the electricity used in the US.[7] Some urban sanitation districts, including the East Bay Municipal Utility District (EBMUD) on San Francisco Bay, have been able to radically reduce their carbon footprint by harvesting biogas energy from decaying restaurant waste and other sources. EBMUD has gone beyond net zero energy consumption and is now selling energy back to the grid. Unfortunately, these biogas systems are not cost-effective in small cities like Arcata.

In the end, city staff worked out a compromise plan. Arcata's sewage treatment plant will be upgraded by investing in and improving elements of the original system—the oxidation ponds and treatment wetlands—as well as building a single oxidation ditch. This hybrid of natural and conventional treatment should keep the effluent within permitted water quality limits even during heavy winter rains.

A single oxidation ditch won't have the capacity to treat all the city's sewage on its own. By deliberate design, the ponds and treatment wetlands will remain an essential part of the system. But space will be left for a second oxidation ditch, which may be needed if rising bay waters drown the ponds and treatment wetlands in the future.

Once an oxidation ditch is running, it must be kept going, or the population of aerobic microbes that do the work will die off. This means the oxidation ditch will have to be powered throughout its life—but operators will be able to turn the aerators down in times of low sewage flow. "We'd be crazy," says Andre, "not to turn it down as far as possible."

Energy consumption at the sewage treatment plant will climb dramatically—and not only because of the power needed to run the oxidation ditch. UV disinfection uses plenty of electricity, and so will the aerators added to the old oxidation pond. As the city staff saw it, the only practical alternative would have been to keep chlorine disinfection, with its toxic hazards, indefinitely. The city council decided against that.

Arcata has begun installing solar panels to transform the sunlight that strikes the open-air treatment system into electricity. The city plans to continue adding panels and selling energy back to the grid. This will put a significant dent in the increased cost of powering the hybrid treatment system planned for the future.

Our compulsive burning of fossil fuels will eventually cause the marshes that inspired this book to vanish under the tides of Humboldt Bay. All of us push the waters higher—when we drive our cars, when we heat our houses, when we flush our toilets. While Arcata's wastewater treatment system is unique, its dilemma is not. Wastewater treatment plants around the world have been built at the edges of estuaries and oceans. Now, many of them have come perilously close to being overwhelmed by rising seas.

The San Francisco Bay area leads the charge in addressing this problem—and a key part of the strategy there is restoration of tidal marshes. Like other wetlands, they provide crucial habitat and filter pollutants out of water. They absorb the shock of powerful waves, buffering the shore from flooding and erosion. They capture and store carbon, helping to offset humankind's fossil fuel addiction. Before we diked and drained and paved them over, tidal wetlands buffered the land from the sea. In the decades to come, they offer the best chance of preserving roads, buildings, airports, and wastewater treatment plants built near shorelines.

NOTES

1. Carollo Engineers (2016). "City of Arcata wastewater treatment facility improvements project: facility plan update and addendum."
2. Korber, D. (October 10, 1999). "Water polluters hard to control." *Long Beach Press-Telegram*, p. A1.
3. Igrejas, A. (June 18, 1998). "California needs to come clean in year of oceans." *Daily News of Los Angeles*, p. N19.
4. Swanson, C.R. (2015). "Annual and seasonal dissolved inorganic nutrient budgets for Humboldt Bay with implications for wastewater dischargers." Thesis, Humboldt State University.

5. Carollo Engineers (2017). City of Arcata wastewater treatment facility improvements project: Facility plan. https://www.cityofarcata.org/DocumentCenter/View/6272
6. Laird, A. (2015). "Humboldt Bay: sea level rise adaptation planning project, phase II report."
7. Gies, E. (March 27, 2017). "How new technologies are shrinking wastewater's hefty carbon footprint." *Ensia*.

13

The Tide Rises

David Sedlak, an environmental engineering professor at the University of California–Berkeley, stands on a levee near San Francisco Bay's eastern shore. Manmade embankments extend for many miles, lining much of the bay's edge, but Sedlak, a lean, intense guy, is fired up about this newly built one. Instead of the usual barren concrete, the bayward face of the levee slopes gently beneath a dense growth of native wetland plants. From muddy clumps of roots and rhizomes placed here only a year ago, the plants have sprouted into a lush palette of green, from the deep dark of Baltic rush to the bright tones of creeping wild rye.

Sedlak is part of a bold experiment. If it succeeds, the project may reshape the East Bay shoreline, restoring a vast acreage of lost tidal wetlands that will be nourished by treated wastewater. The hope is that vegetated levees (the official moniker for the concept is the Horizontal Levee) will save money and energy, recycle treated sewage to create habitat, and help the urbanized East Bay adapt to rising sea levels.

Conventional levees form steep concrete or earthen walls that armor roads and buildings against the bay's powerful waves. The Horizontal Levee is a lovely contrast, a compressed version of a natural habitat long missing from the shoreline. The transition zones, or ecotones, between land and bay were biologically rich places that once hosted a diversity of native plants and animals. Since the Bay Area was settled, wetlands have been diked off from both the open bay and the surrounding land. Between 1800 and 1998, 92 percent of tidal marshes were lost to diking and filling.[1]

"In San Francisco Bay, we've separated the contacts between the terrestrial and the tidal," explains Peter Baye, a consulting ecologist whose deep knowledge of remnant natural wetlands acts as guideline for the creation of the Horizontal Levee. Habitats that once formed a continuous gradient from dry land to salt marsh have been boxed off, separated by dikes. The disappearance of what ecologists call the "back end" of tidal marshes has been a significant loss. Connection with the land once nourished brackish marshes with fresh water and sediment. The habitat created there provided cover for many marsh creatures, including the endangered salt marsh harvest mouse and Ridgway's rail, when tides rose and pushed them out of the flat marsh plain.

Today, restoring both salt marshes and the ecotones that connect them to the land is not just a wildlife conservation technique. It's part of a larger plan to use natural ecosystems to protect the bustling urban zones of the Bay Area from rising waters. As measured by a gauge beneath the Golden Gate Bridge, the bay's level has increased by seven inches from 1900 to 2000. Climate forecasts predict that the rise in bay waters will accelerate in the mid-twenty-first century, climbing an additional fourteen inches by 2050,[2] and continuing to rise to fifty-five inches above current levels by 2100. The greatest immediate threat is from storms that coincide with high tides. Such storm events regularly push water over the top of existing levees, flooding roads.

As a wave crosses a salt marsh, its energy dissipates among the plants and mudflat. Reviving the bay's lost tidal marshes can create a resilient, low-maintenance barrier against storm surges. Rebuilding the back end of tidal wetlands adds another protective barrier, as well as creating a space where salt marshes can migrate inland as the bay level continues to rise.

The notion of the horizontal levee comes from Baye's insights into pre-settlement ecotones, which were fed by seasonal seeps of fresh water that allowed native plants to flourish. Now, levees block any fresh water that runs off during rainstorms—and the majority of the available fresh water is in the form of treated sewage effluent, which is piped past any potential restored wetland areas and into the deep waters of the bay. These deep-water discharges were required by the San Francisco Regional Water Quality Control Board starting in the 1970s.

Baye suggested the idea of redirecting wastewater flows so that treated effluent would filter through a restored land–marsh transition zone. This would have multiple benefits: A restored ecotone would provide plant and wildlife habitat, pump nitrogen out of the water and into the air through the action of denitrifying microbes, and build a layer of peat that would buffer the impacts of rising bay waters. The East Bay Discharge Authority, which pipes effluent from several treatment plants in the area to the bay, would be spared the expense of replacing its aging deep-water pipe, now corroded from decades in salt water.

The concept was embraced by geomorphologist Jeremy Lowe, now with the San Francisco Estuary Institute, who sees it as an inspired way to combine goals and funding. Wastewater managers have access to different sources of money than wetlands restoration advocates, so having both groups behind a project is a big advantage. For Lowe, the creation of the prototype Horizontal Levee at Oro Loma is an important first step on a long road of testing ways to restore wetlands while protecting the bay's shoreline.

"We've been waving our arms around for a long time, but up to now we haven't built anything," he says. "I see it as an experiment to provide scientific information, but also a way to focus the conversation about how we are going to adapt to sea level rise."

The experimental levee, which stands adjacent to the sewage treatment plant at Oro Loma, holds a series of parallel wetland cells. Rushes and sedges send their roots down into carefully prepared layers of substrate. Beneath a shelf of soil lies a layer of gravel mixed with wood chips. As wastewater flows through this underground layer, anaerobic bacteria will transform its load of nitrate into N_2 gas, which dissipates into the atmosphere. This denitrification process takes an ample supply of carbon, which is why wood chips were added to the experimental wetlands. Sedlak hopes that within about ten years, the wetlands will have built up a layer of peat containing enough carbon to fuel denitrification for the long term.

The levee's wet meadows are lush, dense with an assortment of native plants. So far, there's been no significant problem with the invasive weeds common on much of the shoreline. "If we let the native flora reconnect to fresh water sources, that can take care of the weed problems," explains Baye. The theory is that native marsh plants can outcompete invasives when conditions are right—and a little over a year into the life of the Horizontal Levee, that notion is proving true. When I visited in March 2017, the levee was watered by abundant rainfall from the wet winter of 2016–2017. The big question is how the plant community will respond when treated wastewater starts flowing through the system in May.

For now, the water that flows out from the experimental system is routed back to the wastewater treatment plant to be discharged through the long pipe that extends into the bay—because the experiment is too new for regulators to allow actual discharge of effluent via shoreline wetlands. The hope is that if the Horizontal Levee concept can be shown to work, regulators with the regional board will be willing to trade the expensive deep-water discharge pipes of the past half-century for a series of restored shoreline habitats. The idea makes sense for a number of reasons: Throughout the bay, deep-water pipes will be increasingly susceptible to damage as the sea level rises. The long, plant-covered slope of the horizontal levee will provide a far more resilient protection against storm surges than conventional cement or earthen dikes. As time passes, the region's wastewater treatment plants, almost all of them built near the shoreline, will be in growing need of protection from the rising waves.

The Horizontal Levee, with its focus on bringing back lost ecotones, is a departure from decades of wetland restoration practices in San Francisco Bay. Some of California's earliest tidal restoration projects took place in the bay. These efforts were motivated by Section 404 of the Clean Water Act, which requires that developers who destroyed wetlands must restore habitat elsewhere in mitigation.[3] They focused only on restoring salt marsh, the realm of cordgrass and pickleweed. Transition zones were not part of the calculus.

Phyllis Faber, then a consulting biologist, pioneered salt marsh restoration in the bay with the Muzzi Marsh project, begun in 1976. Construction of the Larkspur ferry landing in Marin County required dredging of Corte Madera Creek, an action that destroyed some wetland. In mitigation, Faber designed a restoration of the Muzzi property, which had been diked and drained in the 1950s, and used as a dumping ground for dredge spoils.

Faber's initial restoration plan was simple: Breach the dikes in a few spots to allow the tide to flow in and out. In the first year of tidal flow, she remembers the soil softening. By the second year, native pickleweed had begun to establish. "We were thrilled with how quickly the pickleweed came back," Faber recalls. To accelerate the regrowth of native salt marsh plants, in 1979 a network of channels was dug to improve circulation of tidal waters through the marsh. Later, dredge spoils were deliberately placed at the marsh's edge to provide sediment.[4] A supply of inorganic sediment is an essential ingredient for the establishment of healthy salt marsh; the accumulation of silt allows the marsh to build elevation, necessary for the growth of cordgrass and other native plants. This process has always been part of marsh formation. Over the past six thousand years, salt marsh has established and spread out when bay levels were low and sediment supply was high, then receded again when waters rose.

Faber spent twenty years monitoring the progress of Muzzi Marsh. She tracked subtle changes in the marsh's elevation, and corresponding shifts in plant growth. Pickleweed dominates at higher elevations, cordgrass in the areas that are more deeply flooded with the shifting tides. Now, like other restored and natural salt marshes in the north bay, Muzzi Marsh is eroding away. "When the tide comes in it brings sediment, and when it goes out it carries sediment away," she explains. As waters in the north bay have cleared of suspended sediment, they carry away more than they deliver. "As a result," says Faber, "our marshes are all threatened by the normal wear and tear of tidal action."

The bay now eats away at the earthen dike that was first built to cut the land off from the bay in the 1950s, and that has protected Muzzi Marsh since its restoration (Fig. 13.1). If the old levee collapses, the marsh will be drowned and will become one more patch of open water on the Marin County shore.

"I'm going to be ninety, so I'm not going to be around to offer my opinion, and I hope somebody's watching this." says Faber. If Muzzi Marsh is drowned, a part of her legacy will vanish. Once the marshland floods, the tides are likely to flood buildings that stand behind it—an outcome more likely to grab the attention of officials in the town of Corte Madera.

Sediment supply is a major issue confronting efforts to restore salt marsh in San Francisco Bay. The California Gold Rush reshaped the bay in multiple ways. One major influence was an intense pulse of sediment released by hydraulic gold mining in the Sierra Nevada during the mid-1800s. Whole mountainsides were ripped apart, and their remains eroded into the creeks and rivers of the Sierra, flowing into the Sacramento–San Joaquin Delta and out to San Francisco Bay. New salt marshes developed, often outside the levees that had separated prehistoric wetlands from the tides. At the few ancient, intact salt marshes that survive on the bay, a fringe of newer marsh appeared, fed by the heavy loads of Gold Rush silt.[5] As dams were built to hold back the Sierra's rivers, sediment was trapped behind them and the amounts flowing to the bay declined. Bay waters cleared, and a 1999 analysis found that levels of suspended sediment would continue to decline.[6]

Federal and state regulations forbid the dumping of sediments on wetlands, part of an effort to protect the bay from a wave of filling and development that threatened its survival in the 1960s. Up until 1965, when a band of local activists triggered the establishment of the Bay Conservation and Development Commission (BCDC), the bay was routinely used as a dumping ground for trash and untreated wastewater.

Figure 13.1 Erosion of the levee surrounding Muzzi Marsh, Corte Madera, California. As upstream dams capture sediment and sea level rises, San Francisco Bay's surviving salt marshes are threatened. Photo by Sharon Levy.

Now, the end of the Gold Rush sediment pulse portends a new era for wetland restoration in the region, in which sediment, long disposed of in deep waters to keep it as far from sensitive wetlands as possible, will become a vital resource for marsh restoration.[7]

During her time as a restoration biologist, Faber emphasized that native marsh plants would revive from seed banks surviving in bay soils. Given enough time—sometimes several years—the plants did come back on their own. Limiting human interference seemed best, especially after the US Army Corps of Engineers introduced an Atlantic Coast species of cordgrass, *Spartina alterniflora*, which turned invasive, outcompeting the native Pacific cordgrass, *Spartina foliosa*.

Dense stands of Atlantic cordgrass took over marshes, coopting habitat for an array of native plants as well as birds and their prey. A decade-long effort coordinated by the California Coastal Conservancy used herbicide sprays along with volunteer labor to eradicate the invader. By 2016, Atlantic cordgrass had been removed from most of San Francisco Bay, at a cost of more than $30 million.[8] The lesson that restoration ecologists must be careful what they plant has been learned the hard way.

Yet those working to revive the bay's tidal wetlands now feel an unparalleled sense of urgency. Restored marshes may now need the boost of humans intervening to start native plant growth, especially the freshwater and brackish meadow plants that live in the transition from land to salt marsh. (In these areas, unplanted earth can be quickly taken over by weedy invasives.) The bay's fate is tied to the rate of glacier melt in Greenland and Antarctica, which is accelerating. "If you follow the news over the last year, everything is happening a lot sooner and faster than expected," says Baye. "So we might see that spike in global sea level rise sooner than the expected date of 2050."

The South Bay Salt Pond Restoration Project, one of the largest wetland recovery efforts in the world, is a testing ground for new strategies. The project includes more than sixteen thousand acres of salt ponds, areas that had been diked off from the bay and replumbed to concentrate salt that was harvested by the Cargill company. When Cargill consolidated its salt works, it sold many of its abandoned ponds to the state. The goal now is to restore most of that area to salt marsh, while recreating the lost ecotone where the marsh meets the land.

John Bourgeois, executive project manager with the California Coastal Conservancy, is in charge. He's in a race to capture as much sediment as he can to raise the elevation of the former salt ponds. At Alviso, in the far southern reaches of the bay, groundwater extraction has caused the pond floors to sink by as much as eight feet. For now, the waters of the south bay still carry substantial amounts of suspended sediment. In the north bay, the water has cleared and the dissipation of Gold Rush–era silt is more obvious. It's only a matter of time before sediment depletion comes to the south bay too.

The levee surrounding one of the former salt ponds at Eden Landing Reserve, where there's been relatively little subsidence, was breached in 2006. As the tides flowed in and out, they deposited sediment that built lost elevation, allowing native cordgrass to flourish.

One great advantage is that the salt ponds have retained the sinuous tidal channels formed in ancient marshes before they were diked off from the bay (Fig. 13.2). In long-established marshes, these channels form an intricate pattern of dark water interlaced with the greenery of salt marsh plants. The channels are an integral part of a healthy marsh. They carry tidal waters deep into the habitat, deposit sediment, and build ridges at the channel edges where wildlife shelter among stands of cordgrass. One species particularly dependent on the channel system is the endangered Ridgway's rail, which forages and nests along channel edges (Fig. 13.3). Within eight years of the first dikes being breached at Eden Landing, a pair of nesting Ridgway's rails was found. Their population has grown steadily.

To build elevation in the more subsided areas of the restoration project, Bourgeois is seeking permits to pipe in a slurry of sediments dug up during dredging at local harbors. This will require a floating offloader to set up in the bay. Because of regulations governing the dumping of sediments in the bay, this kind of work must be done before dikes are breached. Getting the permits to add sediment to a diked-off salt pond is a challenge: It's just not possible to obtain a permit to add sediment to habitat that's already reconnected to the bay.

Bourgeois hopes to restore complete marshes, which will mean reviving the ecotones between dry land and wetland long ago lost to diking and urbanization. "We're going to have to do that in a somewhat artificial manner," he says, "because

Figure 13.2 Aerial view of tidal channels in a San Francisco Bay salt pond. Meandering channels can take many years to develop, and are critical to the health of salt marshes. Aerial Archives/Alamy Stock Photo.

Figure 13.3 The Ridgway's rail, an endangered species now breeding in parts of the South Bay Salt Pond Restoration Project. Photo by Leslie Scopes Anderson.

this isn't the north bay where the shoreline is adjacent to rolling hills and farm lands." There's little space between the old salt ponds and the nearest buildings. So the plan is to recreate the lost transition zone on a manmade slope, like the one at Oro Loma's Horizontal Levee.

That will mean using a tremendous amount of fill—hundreds of thousands of cubic yards of it, thousands of dump-truck loads. Bourgeois needs a lot of spare earth to create his transition zones, and that can be an expensive proposition. His plan, which he has tested before during a restoration project at Bair Island, is to use dirt dug up at construction sites. Dumping the soil at Eden Landing will be cheaper for construction firms than taking it to a landfill, which is normal procedure. A dirt broker acts as a middleman, testing the earth to make sure it's not tainted with toxins. The restoration project ends up getting its fill for free. The tactic fits right in with the restoration ecologists' ethic of recycling: old salt ponds into healthy marshes, treated sewage into a water source for wet meadows.

Bourgeois has submitted applications for a series of permits required to create the ecotone slopes he hopes will someday back up restored marshes at Eden Landing's former salt ponds. "We're going to be a huge guinea pig for this process," he says. "I'm proposing several hundred thousand cubic yards of material to be placed in the bay. So it's going to be a wait and see whether or not the agencies can adapt."

Staffers at agencies like the San Francisco Regional Water Quality Control Board and the BCDC understand that circumstances have changed dramatically since regulations were established to protect the bay in the 1960s and 1970s. Adapting to those changes remains a work in progress. "We've been in discussions on our policies for sea level rise and sediment decline issues for two years," says Brenda Goeden, sediment project manager for BCDC. So far, the most radical new permit requests are Bourgeois' petitions to build a new back end onto a series of restored saltmarshes.

Even if by some miracle we halt our combustion of fossil fuels, the process of sea level rise has already been set in motion. Restored tidal wetlands will be an essential part of adapting. Yet wetlands, both natural and restored, coastal and inland, combat global warming in another important way: They lock up carbon that would otherwise be released to the atmosphere as the greenhouse gas carbon dioxide (CO_2).

Wetland plants—bulrush, cattail, or willow in a freshwater wetland, cordgrass or pickleweed in salt marsh—sink their roots into underwater sediments. There's little or no oxygen in these marsh soils, so decomposition is slow. Over time, old roots die off and become incorporated into a layer of peat, rich in carbon, that can be stored for centuries. New generations of plants thrive, sinking their roots down into the peat, building it further.

When wetlands are drained, peat soils that may have accumulated over millennia are exposed to the air. The organic peat soils oxidize, releasing both carbon dioxide and methane (CH_4), a far more potent greenhouse gas, to the atmosphere. Drained wetlands can hemorrhage carbon stores accumulated over thousands of years. The process begins with a rapid burst of carbon emissions, but can continue at lower rates for more than a century.[9]

Wetlands cover only 5 to 8 percent of the planet's land surface but hold 20 to 30 percent of the soil carbon on Earth.[10] A recent analysis found that wetlands in the continental US contain 11.52 Petagrams (Pg, equivalent to 1 billion metric tons) of carbon. Worldwide, wetlands store an estimated 300 to 700 Pg of carbon.[11] For context, all of the living plants on Earth are estimated to store 56 Pg of carbon; the modern atmosphere, pumped up with greenhouse gases from burning fossil fuels, holds 750 Pg; the planet's soils hold 1,500 Pg.[12]

Given these facts, the notion of restoring wetlands would seem to be a no-brainer. There's been debate among scientists, however, because despite their ability to store prodigious amounts of carbon, wetlands are also natural sources of greenhouse gases, especially methane, which the International Panel on Climate Change has determined possesses twenty-five times the global warming potential of carbon dioxide.[13] Wetlands have been identified as the source of about a quarter of the total methane in Earth's atmosphere, and their emissions are expected to rise as the climate warms.[14] Some researchers have argued that while existing wetlands should be protected for their value as wildlife habitat, water quality filters, and flood protection, restoring more wetlands would be irresponsible in the age of global warming.[15]

Bill Mitsch, the wetlands expert and dreamer of landscape-level restoration, strongly disagrees. He and his coworkers analyzed carbon flux data from twenty-one wetlands in the tropical and temperate zones. They found that on average, the amount of methane released is equal to 14 percent of the carbon stored in a wetland's peat layer.[16] This puts the ratio of carbon sequestration to methane release at 19:5. Because the result is lower than the 25:1 methane:carbon dioxide global warming ratio established by the International Panel on Climate Change, this casts wetlands as a net source of greenhouse gas pollution.

When time is added to the analysis, however, that initial conclusion is reversed. Methane decays in the atmosphere, morphing into carbon dioxide through the process of oxidation within eight to ten years.[17] Carbon in the peat of healthy wetlands can remain locked away for millennia. Factor the atmospheric decay of methane into the equation, and long-lived wetlands—those that endure for a century or more—become net sinks of greenhouse gas. The team found that the carbon sequestration is highest in the temperate zone; tropical wetlands release more methane because decomposition happens faster in the heat, and boreal ones sequester less carbon because the Arctic growing season is short.

Methane is generated as anaerobic bacteria break down organic matter in peat. The lack of oxygen in the saturated soils of wetlands, the same factor that allows so much carbon to be stored there, makes methane production possible. For Mitsch, methane emissions are a natural side effect of the processes that make wetlands so important—and no reason to abandon the idea of restoration. In some of his studies, even young, newly restored wetlands have sequestered enough carbon to counterbalance their methane emissions.

Mitsch points out that the carbon in fossil fuels comes from ancient wetland peats, transformed over geologic time into coal, oil, or natural gas. It makes sense, he says, that wetlands are likely the most effective ecosystems on the planet for sequestering carbon.

While Mitsch has focused his work on inland, freshwater wetlands, that principle may hold especially true for tidal wetlands. In the past few years there has been a burst of interest in conserving and restoring coastal marshes, mangrove swamps, and seagrass meadows. These habitats all sequester significant amounts of carbon, and they have been disappearing fast: Estimates of the cumulative global loss of all three habitat types range from 25 to 50 percent over the last fifty to one hundred years.[18] Conversion and destruction of coastal ecosystems has been estimated to release up to 1 billion tons of carbon dioxide every year.

The freshwater and brackish meadows and marshes at the back end of the tidal zone grow and store carbon fast. In the salt marsh flat, plants face the challenge of pumping excess salt out of their systems, so their growth is slower. In terms of carbon storage, both ends of the spectrum are important. Methane is not produced in salt marsh, because salt water is rich in sulfate, which kills off the microbes that release methane as a product of decomposition in freshwater wetlands.

Studies of what's become known as "blue carbon"—the carbon stored by coastal ecosystems—suggests that these habitats store carbon with far greater efficiency than terrestrial forests or prairies. Tropical forests, the justifiable focus of intense conservation efforts and a significant part of the global market in carbon credits, store about 2.5 metric tons of carbon per square kilometer per year. Salt marshes, seagrass beds, and mangroves capture and sequester an estimated sixty to 210 tons of carbon per square kilometer per year.[19] The loss of these habitats has meant a failure to capture 434 billion tons of carbon over the past century.

Reviving these coastal habitats could make a significant dent in our carbon pollution problem, but that will take what marine ecologist Andrew Irving has called "industrial-scale restorations."[20] That's the kind of thing being attempted by the South Bay Salt Pond Restoration Project, with its sixteen thousand acres of existing former salt ponds and ambitions to create more acreage using fill to build ecotone slopes.

To understand the role of wetlands in a world overloaded with human-generated nutrients in water and air, one needs to consider time. Swamps and marshes were here long before we were, when our ancestors were small, timid primates scurrying through swamps we would later dig up and burn. The great wetlands of North America began to form thirteen thousand years ago, when glaciers receded at the end of the Ice Ages and left behind the Great Lakes, the Chesapeake Bay, and the grand delta of central California flowing into San Francisco Bay. Along the edges of these waters, wetlands grew up and socked away vast stores of carbon. In only two hundred years or so, we humans have managed to rip apart and drain more than half of these ecosystems. We started the destruction using our own blood and sweat. Later we used machines fueled by the energy-rich remains of ancient wetlands.

The wetlands of the future must serve human ends, and bring nature into a manmade world to help mop up some of the mess we have made. We can't revive the vast wilderness of the Great Black Swamp, but strategically placed wetlands built into the Midwestern farmscape can do part of its job. The endless tule marshes of prehistoric California are lost, but a series of soggy meadows engineered into the edges of San Francisco Bay's urban sprawl can help the city and its people to thrive.

Out of sight, in the boggy soil and among the roots of the living plants, nitrogen and phosphorus will be absorbed, carbon locked away. Above, rushes and sedges will wave in the wind, their narrow, bright leaves gleaming in the sun. The breeze will carry the smell of wetlands: a combination of thriving growth and gradual rot. From the reeds will come the trill of a marsh wren, the splash of a frog, or the scurry of a mouse.

Enough obstacles stand in the way to make this seem an impossible dream. That's what the powers that be told Arcata's upstarts forty years ago—and they turned out to be wrong. This, too, is a wetland dream worth chasing.

NOTES

1. Lowe, Jeremy, Bob Battalio, Matt Brennan, et al. (2013). "Analysis of the costs and benefits of using tidal marsh restoration as a sea level rise adaptation strategy in San Francisco Bay." *Prepared for the Bay Institute.* http://climate.calcommons.org/sites/default/files/FINAL%20D211228%20Cost%20and%20Benefits%20of%20Marshes%20022213.pdf.
2. Ibid.
3. Peterson, I. (December 31, 1989). "Compensating for filling in a wetland." *New York Times.*
4. Goodwin, P., P.B. Williams (1992). "Restoring coastal wetlands: the Californian experience." *Water and Environmenntal Management* **6**(6): 709–719.
5. Baye, P.R. (2012). "Tidal marsh vegetation of China Camp, San Pablo Bay, California." *San Francisco Estuary and Watershed Science* **10**(2). https://escholarship.org/uc/item/9r9527d7
6. Schoellhamer, D. (2011). "Sudden clearing of estuarine waters upon crossing the threshold from transport to supply regulation of sediment transport as an erodible sediment pool is depleted: San Francisco Bay, 1999." *Estuaries and Coasts* **34**: 885–899.
7. Ibid.; "The Baylands and climate change: what we can do." *Baylands Ecosystem Habitat Goals Science Update 2015, prepared by the San Francisco Bay Area Wetlands Ecosystem Goals Project.*
8. Rogers, P. (April 17, 2016). "San Francisco Bay: massive effort to remove aquatic invader nearly finished." *San Jose Mercury News.*
9. Callaway, J., Steve Crooks (2015). "Science Foundation Chapter 6: Carbon sequestration and greenhouse gases in the baylands." In *Baylands Ecosystem Habitat Goals Science Update 2015, prepared by the San Francisco Bay Area Wetlands Ecosystem Goals Project.*

10. Nahlik, A.M., M.S. Fennessy (2016). "Carbon storage in US wetlands." *Nature Communications* 7(13835). https://www.nature.com/articles/ncomms13835.pdf
11. Lenart, M. (2009). "An unseen carbon sink." *Nature* **3**: 137–138.
12. University of New Hampshire. "GLOBE carbon cycle" (http://globecarboncycle.unh.edu/CarbonPoolsFluxes.shtml).
13. Mitsch, W., Blanca Bernal, Amanda Nahlik, Ulo Mander, et al. (2013). "Wetlands, carbon, and climate change." *Landscape Ecology* **28**: 583–597.
14. Bloom, A.A., Paul Palmer, Annemarie Fraser, David Reay, Christian Frankenberg (2010). "Large-scale controls of methanogenesis inferred from methane and gravity spaceborne data." *Science* **327**: 322–325.
15. Bridgham, S., J.P. Megonigal, J.K. Keller, N.B. Bliss, C. Trettin (2006). "The carbon balance of North American wetlands." *Wetlands* **26**: 889–916.
16. Mitsch, W., Blanca Bernal, Amanda Nahlik, Ulo Mander, et al. (2013). "Wetlands, carbon, and climate change." *Landscape Ecology* **28**: 583–597.
17. Ibid.
18. Pendleton, L., Daniel Donato, Brian Murray, et al. (2012). "Estimating global "blue carbon" emissions from conversion and degradation of vegetated coastal ecosystems." *PLoS One* **7**(9): e43542.
19. Irving, A., Sean Connell, Bayden Russell (2011). "Restoring coastal plants to improve global carbon storage: reaping what we sow." *PLoS One* **6**(3): e18311.
20. Ibid.

14

The Fight This Time

Forty-five years after the passage of the Clean Water Act (CWA), water pollution remains a profound problem. More than forty-seven thousand US waters are impaired. At the rate these lakes, rivers, and estuaries are being cleaned up, it will take more than five hundred years to make them all safe for swimming and fishing.[1] Oliver Houck, a professor of law at Tulane University who has focused on environmental protection since the 1970s, sums up the situation: "We have not had clean water in America," he writes, "in the lifetime of anyone living."[2]

The major source of pollution in the waters of the US, as in other developed countries, is now runoff from farm fields and city streets. These nonpoint sources remain difficult to control.[3] More than 75 percent of the rivers and lakes that fail to meet water quality standards are tainted by nonpoint sources.[4] In terms of nutrient pollution, agricultural runoff is by far the dominant source,[5] triggering harmful algal blooms from Chesapeake Bay to Puget Sound.

The CWA of 1972 addressed point sources of pollution in a decisive and radical way. Section 402 of the CWA applies effluent standards based on the best available treatment technology to city sewage and industrial wastewaters, and puts regulatory power in the hands of the federal Environmental Protection Agency (EPA). Regulation under this scheme has brought dramatic improvement in water quality. Before the CWA was enacted, major urban river systems throughout the country had such low levels of dissolved oxygen that fish kills became routine,[6] and urban beaches were often closed due to fecal contamination. By the late 1990s, dissolved oxygen levels had improved in about 70 percent of river reaches and watersheds studied by the EPA, and fish had returned to many waters. Beach closures decreased. Problems remain, especially in cities like Chicago and Baltimore, where heavy rains can overwhelm treatment systems, releasing raw sewage downstream. Still, in terms of curbing point source pollution, the CWA has made a critical difference.

The rise of pollution from unregulated nonpoint sources has eaten away at these water quality gains. The Mississippi River basin, whose waters flow into the northern Gulf of Mexico, may be the most dramatic example. In August 2017, the Gulf's dead zone grew to an unprecedented 8,776 square miles, about the size of New Jersey.[7] Three-fourths of the nutrient load triggering the Gulf's vast dead zone flows from farms above the confluence of the Ohio and Mississippi Rivers, more than 975 river miles upstream.[8] Heavy loads of nitrate taint drinking water sources in the basin. The pattern is repeated throughout the US: Wisconsin's lakes are polluted by runoff from dairy farms, Washington's rivers and estuaries by agricultural fertilizers, Maryland's by industrial poultry farming.

Before 1972, water quality regulation was left to the states. The federal CWA was written because state programs had failed.[9] Local dischargers held enormous political clout, enough to shape the water quality standards in each state. This created a "race to the bottom" in which states lowered the bar for polluters in order to lure industries across state lines. The legislators who crafted the CWA were keenly aware of these problems. Senator Edmund Muskie, a driving force behind the CWA, wanted to transfer environmental protection entirely to the federal government and opposed including state water programs at all.[10]

In 1972, solutions to the problem of nonpoint pollution were unclear and politically daunting. Answers then, as now, involved putting some limits on land use, traditionally viewed as the province of the states. State officials lobbied hard to keep control. In the end, the framers of the CWA let that tradition stand, but they did give the federal EPA authority to monitor state actions, and to step in when states fail to act.

In the decades since, nonpoint source pollution has grown worse. Most states have only voluntary programs to address nonpoint pollution. The industries that would be affected by stronger regulation have resisted any move toward enforceable pollution limits.[11] In Congress and the courts, the struggle for control of nonpoint pollution has moved at a slow creep. The original CWA called for water pollution control planning by each state. Planning commenced, but no action followed.

In 1987, Congress amended the CWA, adding Section 319, which requires each state to identify impaired waters and create a plan to clean them up. Senator Robert Stafford, then chairman of the Senate Committee on Environment and Public Works, described Section 319 as "a first step in tackling the problem—a trial run, to see if allowing the States the option to develop a control program will indeed abate nonpoint source pollution across the Nation."[12] The states have been reluctant to move, and Congress failed to back the program up with adequate funding. The trial run has failed.

Grappling with nonpoint pollution is not a task anyone in government has been eager to take on. It took a series of citizen lawsuits in the 1990s to force the EPA to acknowledge its responsibility to step in when states fail to act. The agency has been slow to push the issue, and prefers to work in cooperation with the states, despite decades of inertia and continuing lawsuits from environmental groups.

William Andreen, an expert on water law and a professor at the University of Alabama School of Law, has closely tracked this history. "The goals of the Clean Water Act are unlikely to ever be fulfilled unless something other than a voluntary approach is taken to nonpoint source pollution," he writes.[13]

The opposition to such a change is fierce. The problem is not a scientific or technical one. We know how to decrease nutrient pollution, and as Houck has written, it's not rocket science.[14] It is instead a matter of convincing farmers—all of them—to use strategic fertilizer application, vegetated buffers, restored wetlands, and other management practices that limit the amount of nitrogen and phosphorus dumped on the land and leached into rivers. It might mean changing crops, giving up some cropland. From the birth of the CWA, farm runoff has been exempted from regulation—a mark of the agricultural lobby's power. Setting and enforcing water quality standards for farm runoff will take a departure from the longstanding status quo, which has left agriculture above the law.

"The issue is often portrayed as imposing regulation on hapless family farms," says Robin Kundis Craig, professor of law at the University of Utah and author of numerous works on the CWA. "That scenario tugs on an American heartstring. The reality is that many farms in the US are industrial scale, owned by corporations that are just as capable of putting the financial effort into controlling their pollution as any industry in the US." In Craig's view, the main obstacle to dealing with agricultural pollution is politics and an outdated image of American farming.

Controlling pollution takes effort, investment, and a change from business as usual. The responsible parties, whether they're industries or cities or individuals, don't volunteer to do it; they have to be pushed. In the case of agricultural pollution, the dominant strategy in the US has been to offer financial incentives in an effort to lure farmers into using conservation practices. In the handful of states that have instituted regulations on nonpoint farm pollution, there's a demand for clean water that can counterbalance the political leverage of agriculture. In Washington and Oregon, it's the need to protect dwindling salmon runs.[15] Yet even the few states with regulations on the books seldom enforce them.[16] Forty-five years of the current approach hasn't put a real dent in nonpoint pollution.

Alternatives exist, and they have been shown to work in some European nations. In Denmark, agricultural runoff has been a major driver of eutrophication in estuaries and coastal waters. In the late 1980s, a government initiative made farm subsidies contingent on strict nutrient budgeting, which limited farmers to using enough fertilizer to support only 90 percent of their traditional yield (see Chapter 10). New rules were applied to manure storage, the timing of manure application, and wetland restoration. Government auditors monitored fertilizer use and levels of nitrogen and phosphorus in runoff. By 2013, excess nitrogen flowing from the landscape had been cut in half, contributing to marked progress toward recovery in coastal ecosystems.[17]

Using a different approach, the Netherlands succeeded in curbing nutrient pollution by imposing a tax on excess nitrogen and phosphorus runoff from farms. Taxes on pollution are unpopular, however, and this one proved expensive to enforce. Eventually, the Dutch abandoned taxation and sought an alternative way of limiting fertilizer use.[18]

The European Union's Water Framework Directive is similar to the CWA. On both sides of the Atlantic, efforts to control point source pollution have proven far more effective than those aimed at nonpoint sources. In most European countries, nonpoint source pollution controls remain voluntary, as they are in the US.

Part of the problem is a question of scale. In 1972, point source regulation in the US involved tens of thousands of permits. To address nonpoint sources today would involve regulating millions of farms. There's no single technological fix that would work everywhere, so regulation must allow the flexibility to deal with local conditions. A successful program will need to include participation from affected farmers. Many who have studied the nonpoint source conundrum believe that despite these complexities, enforceable regulations are the only road forward.[19] The biggest obstacle is a lack of political will.

Could the Danish or Dutch approaches work in the US? "It could be done," muses Andreen, "but you might have another revolutionary war on your hands . . . I wouldn't want to be the politician to propose that in this country."

The fight to control nonpoint source pollution has often focused on Section 303(d) of the CWA, which requires that states designate a maximum concentration of a given pollutant that can be allowed in affected waters, called a Total Maximum Daily Load (TMDL). If a state fails to do this, the federal EPA must. This provision was written into the original CWA but was largely ignored until a series of citizen lawsuits in the 1990s forced EPA to move. The agency convened a committee with representatives from state water agencies, agribusiness, timber, municipal and industrial point sources, Native Americans, and academics. Their task was to reach a consensus on regulations for a national TMDL program.

That consensus was never reached. "At bottom," writes Houck, "the timber and agriculture industries were not going to accept that CWA Section 303(d) covered nonpoint sources or that, if it did, it required implementation plans . . . A Farm Bureau Federation representative wrote that the program had been 'hijacked by a vast national bureaucracy of parasites.'"[20]

Not much has changed since Houck described the political battle over TMDLs in the late 1990s and early 2000s. Agriculture continues to resist the establishment of numerical limits on pollution from nonpoint sources. The industry favors the narrative standards that have long been used in state regulations—and such standards are nearly impossible to enforce. A narrative standard may state, for example, that phosphorus and nitrogen may not be discharged in amounts that will result in growths of algae that impair waters for their best uses. Such standards leave regulators with the burden of tracking nonpoint pollution back to myriad sources throughout a watershed, and of proving that individual farms or landowners bear responsibility for degraded conditions downstream. In contrast, numerical standards are clear: Like a set speed limit, they leave little room for evasion or argument. Using narrative standards to govern nutrient pollution is like posting highway signs that say "drive safely" instead of listing a legal speed.[21]

In 1998, EPA formally recommended that states accelerate the development and adoption of numerical nutrient water quality standards. The agency pointed out that numerical standards are needed to create measurable baselines for acceptable levels of nitrogen and phosphorus, to allow evaluation of best management practices meant to reduce nonpoint nutrient pollution, and for the development of water quality targets under the TMDL program.

EPA projected that all fifty states would have set numerical standards by 2004. At the time, six states had adopted numerical standards for at least one type of nutrient for one type of water body. By 2008, one more state had joined that list.[22] As of September 2017, only Hawaii had a complete set of numerical criteria for all its state waters. Minnesota, Wisconsin, New Jersey, and Florida had made substantial progress, and had established standards for nitrogen and/or phosphorus on two or more kinds of waters. But twenty-six states had no numerical water quality standards at all.[23] Thirteen of those states were in the Mississippi River watershed, the source of the heavy nutrient loads that have been killing the northern Gulf of Mexico for decades.

The most dramatic sign of progress on nonpoint pollution is in the Chesapeake Bay region, where a faltering restoration effort was boosted by the nation's first regional TMDL program, established in 2010. Water clarity is improving, allowing a rebirth of underwater plants. In 2016, seagrass beds in the bay grew to cover almost 100,000 acres—three times the vegetated expanse in the 1980s. The bay's dead zone is shrinking.[24]

This success has been possible because of federal programs, including the EPA's Chesapeake Bay Program. The Trump administration has proposed to eliminate the program's $73 million in federal funds. It's part of a broad attack on environmental protection, which includes rolling back regulations and slashing EPA's budget by one-third. If the president gets his way, he will end EPA programs that support restoration efforts in the Great Lakes and Puget Sound watersheds, and gut the agency's enforcement and research capabilities.

Trump campaigned on his intention to dismantle the EPA. In a March 2016 debate between Republican presidential candidates, Trump announced that if elected he would eliminate "The Department of Environment Protection," his misnomer for EPA. "We're going to have little tidbits left but we're going to get most of it out," he said.[25]

The man Trump put in charge at the EPA built his career on attacking the agency. As attorney general of Oklahoma, Scott Pruitt signed a legal brief supporting the Farm Bureau Federation's suit against the Chesapeake Bay TMDL. He sued the EPA fourteen times, seeking to halt establishment or enforcement of new regulations. In all but one of these cases, Pruitt's co-litigators—most often energy companies—contributed to Pruitt's campaign or a political action committee affiliated with him. Together with Murray Energy, Peabody Energy, Southern Company, and others, Pruitt worked to stymie the EPA's efforts to regulate air pollution across state boundaries, improve air quality in national parks, reduce emissions of greenhouse gases, and extend CWA protections to headwater streams and wetlands.

In 2011, Pruitt submitted a letter to the EPA accusing the agency of grossly overestimating the amount of air pollution caused by energy companies drilling new natural gas wells in Oklahoma. He failed to mention that the letter had been written by lawyers for Devon Energy, one of Oklahoma's biggest oil and gas companies. Investigative reporting by the *New York Times*[26] documented this and other examples of Pruitt's cozy relationship with industries he was meant to regulate.

Early in his tenure, Pruitt closed down a section of the attorney general's office devoted to environmental protection and established a "federalism unit" to fight regulation by federal agencies. His biography on the Federalist Society website described him as "a leading advocate against the EPA's activist agenda . . . he is leading the charge against EPA's proposed Clean Power Plan and Waters of the United States rules for their unlawful attempt to displace state sovereignty in an environmental regulatory context."[27]

Pruitt has often described the EPA as a rogue agency, reaching beyond its legal authority to enact oppressive regulations. As the agency's administrator, he says he wants to bring the EPA "back to basics," focused on its mission as defined in landmark environmental legislation passed in the 1970s. He told the House Appropriations Subcommittee that his intention as EPA administrator is to "focus on the rule of law. We're reversing an attitude and approach that one can simply reimagine authority under statutes passed by this body . . . Any action by the EPA that exceeds the authority granted by Congress, by definition is inconsistent with the agency's mission."[28]

Pruitt never offers specific examples of the EPA exceeding its authority. His office did not respond to requests for comment. Based on his publicly available remarks, Pruitt's imagined original EPA seems to have been a meek institution, deferring at all times to the power of the states. That was never the intention of those who created the agency—nor was it the reality at EPA in its early years. EPA was designed as a strong advocate and enforcer of pollution control, with the authority to ride herd on state governments accustomed to letting things slide.

President Richard Nixon created the EPA in 1970. He was not an environmentalist, but he was compelled by public demand for an end to flaming rivers, oil spills, and smog-filled skies. Russell Train, who was then chairman of the president's Council on Environmental Quality, shaped the new agency. "What we needed—and what the public wanted—was an organization with a clearly defined mission: to be the sharp, cutting edge of environmental policy in the government," he said.[29] Train would later become EPA's second administrator.

Nixon echoed Train's vision of the EPA in his letter to Congress, when he proposed creating the EPA as a "strong, independent agency" that would view the environment as a "single, interrelated system."[30] The new agency's sole purpose would be environmental protection. Congress accepted the proposal, and in new legislation gave EPA responsibility for pollution control nationwide. The Clean Air Act, passed in 1970, authorized the agency to set air quality standards for the benefit of public health—and barred it from considering the costs to industry.[31] Two years later, the CWA became law, for the first time giving official consideration to the well-being of fish, wildlife, and ecosystems as well as human health.

William Ruckelshaus, EPA's first administrator, had witnessed the absence of state-level environmental enforcement while working at the Indiana State Board of Health. EPA's role was to set national standards, then delegate enforcement back to the states. If state officials proved too weak to curb local polluters, EPA could step in. Ruckelshaus described EPA as the muscular enforcer states could use to compel reluctant industries to clean up.

"Prior to EPA," he remembered, "there was no federal oversight. There was no 'gorilla in the closet.' Absent that, it was very hard to get widespread compliance."[32]

The EPA, along with the Clean Water and the Clean Air Act, was created to provide all US citizens with a healthy environment—regardless of which state or neighborhood they live in. The agency can set national environmental standards based on solid science, and enforce them. The framers of these laws recognized that government at any level could not always be relied on to act as a strong enforcer, so they built redundancy into the pollution control system. EPA and state governments both hold authority. But citizens have the right to sue to goad the government into action—and such citizen lawsuits have moved environmental protection forward many times over the years.

"If Pruitt were committed to getting EPA back to its origins, he'd be acting much differently than he is," says Jon Devine, a senior attorney in the water program at the Natural Resources Defense Council. "His actions as administrator are antithetical to the mission of EPA, and endanger water quality across the country."

One of Pruitt's goals as EPA administrator is to take back the Waters of the United States (WOTUS) rule, written by EPA staff during the Obama administration. The rule clarifies which waters are covered by the CWA, and would extend protections to headwater streams and wetlands that have a significant impact on downstream water quality. WOTUS was based on more than 1,200 peer-reviewed studies and over 1 million public comments.[33]

In June 2017, Pruitt announced that EPA would withdraw the rule. His action followed an executive order in which President Trump instructed the EPA administrator to "rescind or revise" the WOTUS rule. The order tells officials to interpret the reach of the CWA according to the late Justice Antonin Scalia's opinion in the landmark case of *Rapanos v. United States Army Corps of Engineers*. In the *Rapanos* case, the US Supreme Court split down the middle; there was no majority. The court's divided findings in the case have fueled years of argument and confusion, which the WOTUS rule was meant to clarify. Scalia's opinion dismissed reams of scientific research on the tight relationship between wetlands and downstream water quality.[34] In using the Scalia opinion alone, the Trump order discards a large body of legal thought that allows greater leeway for protection of wetlands and small streams.

"The WOTUS rule was well within the agency's legal authority," says Devine. "It was rigorously supported by agency science. By contrast, Pruitt's planned repeal of that rule has no basis in law or science."

Devine notes that there's a strong legal framework backing up the WOTUS rule—and environmental groups can resort to lawsuits to defend it. But even if the rule can be preserved, major damage may be done. "This administration can turn a blind eye to pollution," he says. "Failing to enforce the law is in many cases just as bad as trying to roll it back."

Over the last four decades, bedrock environmental laws have shifted the baseline so that most Americans take their relative safety for granted. The clearest statistics are available for the Clean Air Act. Since its passage, despite a US population increase of more than 50 percent and a 250 percent rise in gross domestic product, the nation has seen a 70 percent reduction in emissions of air pollutants targeted under the law.[35] This translates into illnesses prevented and lives saved: an estimated 160,000 people in 2010. But that trend will hold only as long as the law continues to be enforced.

Great progress has been made since the days of mass typhoid and cholera epidemics in American cities. Water pollution can still threaten human health, however—a fact that's most obvious in places like Des Moines, Iowa, with its ongoing struggle against nitrate contamination, and in Flint, Michigan, with its corroding lead pipes. Climate change will intensify problems related to nonpoint source water pollution. Heavier seasonal rains in the Midwest threaten to increase loads of nutrients, pathogens, and toxic chemicals in farm and urban runoff.[36] In the arid Southwest, droughts lead to intensifying wildfires and increased soil erosion. Both patterns can affect the quality of drinking water.

Pruitt never discusses the possible impacts of slashing environmental enforcement nationwide. In his first address to EPA staff, Pruitt made no mention of human or environmental health. Instead he asserted that "federalism matters," and that "regulators exist to give certainty to those that they regulate."[37]

Over the five decades of its existence, both Republican and Democratic presidents have helped to build the EPA. The only time the agency has experienced an attack comparable to the one unleashed by the Trump Administration was during 1981–1983, in the early years of the Reagan presidency. Reagan nominated Anne Gorsuch as EPA administrator. Her goals were much the same as Pruitt's: to shrink the agency, roll back rules, and redistribute more regulatory power to the states. Her first major organizational move was to break up EPA's office of enforcement. Career EPA staff saw this as a strategy to silence some of the most effective environmentalists at the agency.[38] Gorsuch also recommended deep cuts to EPA's budget. On her watch, enforcement actions taken against polluting industries fell by 79 percent.

A wave of scandal soon engulfed EPA. Gorsuch had promised the oil refiner Thriftway Company not to act on its failure to remove lead from gasoline. Her aide, John Hernandez, allowed Dow Chemical to edit an EPA report on dioxin contamination caused by the company's plant in Midland, Michigan.[39] Rita Lavelle, a political appointee in charge of the $1.6 billion Superfund program, was fired and later convicted of lying to Congress. Gorsuch resigned in March 1983 amid Congressional investigations of EPA's mismanagement of the Superfund program, intended to clean up hazardous waste sites.

The Gorsuch regime was brought down by resistance from Congress, legal and political activism by environmental groups, and public concern. In the aftermath, Reagan asked William Ruckelshaus, who had helped found the agency, to return to the EPA as administrator. He was welcomed back by EPA staff and spent two years working to rebuild the agency's enforcement ability and restore some of the budget cuts. Since he left the agency in 1985, EPA's responsibilities have grown while its funding has stayed the same.

Public outrage, Ruckelshaus has observed, is key to the existence of EPA and other environmental regulators. It's the force that ultimately drives the control of air pollution, the construction of sewage treatment plants, and the restoration of wetlands. Nixon had no sympathy for environmentalists; he called them "crazies."[40] Yet he laid out the blueprint for EPA, and later doubled the agency's budget.

"[Nixon] created EPA for much the same reason Reagan invited me to return to the agency in 1983," Ruckelshaus has said. "Because of public outrage about what was happening to the environment. Not because Nixon *shared* that concern, but because *he didn't have any choice*."[41]

Those of us who study history and care about the future need to keep politicians in that same position: protecting the environment because the voters demand it. The task is unglamorous but essential—like the life of a cattail in Arcata's marsh.

NOTES

1. US Environmental Protection Agency, Office of Wetlands (2011). "A national evaluation of the Clean Water Act Section 319 program."
2. Houck, O. (1999). "TMDLS IV: the final frontier." *Environmental Law Reporter* **29**: 10469–10486.
3. Drevno, A. (2016). "Policy tools for agricultural nonpoint source water pollution control in the US and EU." *Management of Environmental Quality* **27**: 106–123.
4. Andreen, W.L. (2016). "No virtue like necessity: dealing with nonpoint source pollution and environmental flows in the face of climate change." *Virginia Environmental Law Journal* **34**: 255–296.
5. Houck, O. (1999). "TMDLS IV: the final frontier." *Environmental Law Reporter* **29**: 10469–10486.
6. Press, D. (2015). "Failure where there is no pipe." Chapter 4 in "American environmental policy: the failures of compliance, abatement and mitigation," pp. 87–123. Cheltenham, UK and Northampton, Massachusetts: Edward Elgar Publishing.
7. National Oceanic and Atmospheric Administration. (2017). "Gulf of Mexico dead zone is the largest ever measured" (http://www.noaagov/media-release/gulf-of-mexico-dead-zone-is-largest-ever-measured).
8. Houck, O. (1999). "TMDLS IV: the final frontier." *Environmental Law Reporter* **29**: 10469–10486.
9. Houck, O. (2014). Cooperative federalism, nutrients, and the Clean Water Act: three cases revisited. *Environmental Law Reporter* **44**: 10426–10442.
10. Ibid.
11. Ibid.; Andreen, W.L. (2016). "No virtue like necessity: dealing with nonpoint source pollution and environmental flows in the face of climate change." *Virginia Environmental Law Journal* **34**: 255–296.
12. Andreen, W.L. (2016). "No virtue like necessity: dealing with nonpoint source pollution and environmental flows in the face of climate change." *Virginia Environmental Law Journal* **34**: 255–296.
13. Ibid.
14. Houck, O. (2002). "The Clean Water Act TMDL Program V: aftershock and prelude." *Environmental Law Reporter* **32**: 10385–10405.

15. Craig, R.K., Anna M. Roberts (2015). "When will governments regulate nonpoint source pollution? A comparative perspective." *Boston College Environmental Affairs Law Review* **42**: (http://lawdigitalcommons.bc.edu/cgi/viewcontent.cgi?article=2153&context=ealr).
16. Andreen, William. Interview, July 28, 2017.
17. Riemann, B., Jacob Carstensen, Karsten Dahl, et al. (2016). "Recovery of Danish coastal ecosystems after reductions in nutrient loading: a holistic ecosystem approach." *Estuaries and Coasts* **39**: 82–97.
18. Drevno, A. (2016). "Policy tools for agricultural nonpoint source water pollution control in the US and EU." *Management of Environmental Quality* **27**: 106–123.
19. Ibid.; Press, D. (2015). "Failure where there is no pipe." Chapter 4 in "American environmental policy: the failures of compliance, abatement and mitigation," pp. 87–123 Cheltenham, UK and Northampton, Massachusetts: Edward Elgar Publishing.; Andreen, W.L. (2016). "No virtue like necessity: dealing with nonpoint source pollution and environmental flows in the face of climate change." *Virginia Environmental Law Journal* **34**: 255–296.
20. Houck, O. (2002). "The Clean Water Act TMDL Program V: aftershock and prelude." *Environmental Law Reporter* **32**: 10385–10405.
21. Houck, O. (2014). "Cooperative federalism, nutrients, and the Clean Water Act: three cases revisited." *Environmental Law Reporter* **44**: 10426–10442.
22. US Environmental Protection Agency. (2008). "State adoption of numeric nutrient standards, 1998–2008." EPA-821-F-08-007.
23. US Environmental Protection Agency. (2017). "State progress toward developing numeric nutrient water quality criteria for nitrogen and phosphorus" (https://www.epagov/nutrient-policy-data/state-progress-toward-developing-numeric-nutrient-water-quality-criteria).
24. Baye, R. (2017). "State leaders call on feds to keep Chesapeake Bay program funds" (http://wypr.org/post/state-leaders-call-feds-keep-bay-program-funds).
25. Feldscher, K. (March 3, 2016). "Trump says he'd eliminate 'Department of Environment Protection.'" *Washington Examiner.*.
26. Lipton, E. (December 6, 2014). "Energy firms in secretive alliance with attorneys general." *New York Times.*
27. Environmental Data & Governance Initiative. (2017). "Scott Pruitt's first address to the EPA: as annotated by EDGI" (https://envirodatagov.org/scott-pruitts-first-address-epa-annotated-edgi/).
28. C-SPAN (2017). "Fiscal year 2018 EPA budget: EPA administrator Scott Pruitt testified before a House Appropriations Subcommittee" (https://www.c-spanorg/video/?429947-1/epa-administrator-pruitt-testifies-fy-2018-budget).
29. Gorn, M. (1992). "Russell Train oral history interview. EPA History Interview-2" (https://archive.epa.gov/epa/aboutepa/russell-e-train-oral-history-interview.html).
30. Nixon, R. (1970). "Special message to the Congress about reorganization plans to establish the Environmental Protection Agency and the National Oceanic and Atmospheric Administration." *The American Presidency Project* (http://www.presidency.ucsb.edu/ws/?pid=2575).
31. Farber, D. (2017). "The truth about environmental originalism." *Legal Planet* (http://legal-planet.org/2017/06/19/lets-talk-about-environmental-originalism/).

32. Ruckelshaus, W.D. (1992). "Oral history interview. EPA History Interview-1" (https://archive.epa.gov/epa/aboutepa/william-d-ruckelshaus-oral-history-interview.html).
33. Mufson, S., Juliet Eilperin (June 27, 2017). "Trump administration to propose repealing rule giving EPA broad authority over water pollution." *Washington Post.*
34. Rosenbaum, S. (2006). "When public health meets market forces: *Rapanos v. US Army Corps of Engineers.*" *Public Health Reports* **121**: 769–772.
35. Samet, J., Thomas Burke, Bernard Goldstein (2017). "The Trump administration and the environment—heed the science." *New England Journal of Medicine* **376**: 1182–1188.
36. Andreen, W.L. (2016). "No virtue like necessity: dealing with nonpoint source pollution and environmental flows in the face of climate change." *Virginia Environmental Law Journal* **34**: 255–296; Gaffield, S., Robert Goo, Lynn Richards, Richard Jackson (2003). "Public health effects of inadequately managed stormwater runoff." *American Journal of Public Health* **93**: 1527–1533.
37. Environmental Data & Governance Initiative. (2017). "Scott Pruitt's first address to the EPA: as annotated by EDGI" (https://envirodatagov.org/scott-pruitts-first-address-epa-annotated-edgi/).
38. Sellers, C., Lindsey Dillon, Jennifer Liss Ohayon, et al. (2017). "The EPA under siege: Trump's assault in history and testimony." *Environmental Data & Governance Initiative* (https://envirodatagov.org/publication/the-epa-under-siege/).
39. Shabecoff, P. (March 16, 1983). "Scheuer says EPA aide let Dow delete dioxin tie in draft report." *New York Times.*
40. Ruckelshaus, W.D. (1992). "Oral history interview. EPA History Interview-1" (https://archive.epa.gov/epa/aboutepa/william-d-ruckelshaus-oral-history-interview.html).
41. Ibid.

INDEX